## Acclaim for *Rats*

"This is a wonderful book about the despised creatures with whom New Yorkers share their city. Rats have been hunted down here for centuries, but remain unvanquished. As Mr. Sullivan reminds us—in detailed, graceful prose—they are as much part of the city's history as any part of its human alloy. One thing is certain: after reading this book you will understand much more about that history, and never look at a rat in the same way again." **—Pete Hamill, author of *Forever* and *A Drinking Life***

"Eloquent."***—Entertainment Weekly***

"*Rats* is a sort of bizarro-Walden, an exercise in really knowing one small, unremarkable and, in this case, revolting plot of ground. In Sullivan's mirror *Walden* it's the rats, not the people, who live lives of quiet desperation, 'hiding beneath the table of man, under stress, skittering in fear.' "***—Newsday***

"Sullivan leavens his systemic study with anecdotal digressions, approaching his fleet-footed, fast-food-loving quarry with a naturalist's curiosity and a storyteller's fluency."***—New York* magazine**

"*Rats* is a must-read. Don't let this book scurry out of your sight."—***San Antonio Express***

"Hugely entertaining."***—Village Voice***

"Sullivan takes us deeper into the world of rats than most of us, left to ourselves, would be willing to go."***—Dallas Morning News***

"*Rats* will both entertain and edify you about a part of the world you never thought much about."***—Chicago Sun-Times***

"Who knew a book about one of nature's most reviled creatures could make such great bedside reading? I thoroughly enjoyed this historical chronicle of rodents in New York City. It's not only a history of rats, but it is also a social history of the city."—**Book Sense**

"Sullivan beguiles us with remarkable tales about an inexhaustible topic."—***Playboy***

"[A] sublime book."—***Wired***

"*Rats* might generate sympathy for a truce in the longstanding war between rats and the human civilization they mimic."—***Colorado Reporter-Herald***

"Skittering, scurrying, terrific natural history."—***Kirkus Reviews,* starred review**

"This book is a must pickup for every city dweller, even if you'll feel like you need to wash your hands when you put it down."—***Publishers Weekly,* starred review**

# RATS

BY THE SAME AUTHOR

*The Meadowlands*
*A Whale Hunt*

# RATS

*Observations on the History and Habitat of the City's Most Unwanted Inhabitants*

ROBERT SULLIVAN

BLOOMSBURY

Published by Bloomsbury Publishing, New York and London
Distributed to the trade by Holtzbrinck Publishers

All papers used by Bloomsbury Publishing are natural, recyclable products
made from wood grown in well-managed forests. The manufacturing processes
conform to the environmental regulations of the country of origin.

The Library of Congress has cataloged the hardcover edition as follows:

Sullivan, Robert, 1963–
Rats : observations on the history and habitat of the city's most unwanted inhabitants / Robert Sullivan.—1st U.S. ed.
p. cm.
ISBN 1–58234–385–3 (hardcover)
1. Rats—New York (State)—New York—Anecdotes. 2. Urban pests—New York (State)—New York—Anecdotes. 3. Sullivan, Robert, 1963– I. Title.
QL795.R2S85 2003
599.35′21756—dc22
2003016293

First published in the United States by Bloomsbury Publishing in 2004
This paperback edition published in 2005

Paperback ISBN 1-58234-477-9
ISBN-13 9781582344775

3 5 7 9 10 8 6 4 2

Typeset by Hewer Text Ltd, Edinburgh
Printed in the United States of America
by Quebecor World Fairfield

*For Suzanne*

# CONTENTS

You always have to think beyond the structure. Think about what is going on underneath and all around, because that is where the rats are located. The more you look into it, the more you will most likely find.

*—John Murphy, an exterminator, in*
Pest Control Technology *magazine*

I think his fancy for referring everything to the meridian of Concord did not grow out of any ignorance or depreciation of other longitudes or latitudes, but was rather a playful expression of his conviction of the indifferency of all places, and that the best place for each is where he stands. He expressed it once in this wise: "I think nothing is to be hoped from you, if this bit of mould under your feet is not sweeter to you to eat than any other in this world, or in any world."

*—Ralph Waldo Emerson, in a*
*remembrance of Henry David Thoreau*

It avails not, time nor place—distance avails not,
I am with you, you men and women of a
   Generation, or ever so many generations hence,
Just as you feel when you look on the river and sky, so I felt,
Just as any of you is one of a living crowd, I was one of
   a crowd . . .

*—Walt Whitman, "Crossing Brooklyn Ferry"*

*Chapter 1*

# NATURE

WHEN I WROTE the following account of my experiences with rats, I lived in an apartment building on a block filled with other apartment buildings, amidst the approximately eight million people in New York City, and I paid rent to a landlord that I never actually met—though I did meet the superintendent, who was a very nice guy. At this moment, I am living out of the city, away from the masses, in a bucolic little village with about the same number of inhabitants as my former city block. I wouldn't normally delve into my own personal matters, except that when I mention my rat experiences to people, they sometimes think I took extraordinary measures to investigate them, and I didn't. All I did was stand in an alley—a filth-slicked little alley that is about as old as the city and secret the way alleys are secret and yet just a block or two from Wall Street, from Broadway, and from what used to be the World Trade Center. All I did was take a spot next to the trash and wait and watch, rain or no rain, night after night, and always at night, the time when, generally speaking, humans go to sleep and rats come alive.

Why rats? Why rats in an alley? Why anything at all in a place that is, let's face it, so disgusting? One answer is proximity. Rats live in the world precisely where man lives, which is, needless to say, where I live. Rats have conquered every continent that humans have conquered, mostly with the humans' aid, and the not-so-epic-seeming story of rats is close to one version of the epic story of man: when they arrive as immigrants to a newfound land, rats push out the creatures that have preceded them, multiply to such an extent as to stretch resources to the limit, consume their way toward famine—a point at which they decline, until, once again, they are forced to fight, wander, or die. Rats live in man's parallel universe, surviving on the effluvia of human society; they eat our garbage. I think of rats as our mirror species, reversed but similar, thriving or suffering in the very cities where we do the same. If the presence of a grizzly bear is the indicator of the wildness of an area, the range of unsettled habitat, then a rat is an indicator of the presence of man. And yet, despite their situation, rats are ignored or destroyed but rarely studied, disparaged but never described.

I see that I am like one person out alone in the woods when it comes to searching out the sublime as it applies to the rat in the city. Among my guidebooks to nature, there is no mention of the wild rat, and if there is, the humans that write the books call them invaders, despised, abhorred, disgusting—a creature that does not merit its own coffee-table book. Here is the author of a beautiful collection of photographs and prose joyously celebrating the mammals of North America as he writes about rats: "There comes a time when even the most energetic of animal lovers must part ways with the animal kingdom." He goes on: "No matter how much you like animals there is *nothing* good to say about these creatures . . ." It is the very ostracism of the rat, its exclusion from the pantheon of natural wonders, that makes it appealing to me, because it begs the question: who are we to decide what is natural and what is not?

What makes me most interested in rats is what I think of as our common habitat—or the propensity that I share with rats toward areas where no cruise ships go, areas that have been deemed unenjoyable, aesthetically bankrupt, gross or vile. I am speaking of swamps and dumps and dumps that were and still are swamps and dark city basements that are close to the great hidden waters of the earth, waters that often smell or stink. I am speaking, of course, of alleys—or even any place or neighborhood that might have what is commonly referred to as a "rat problem," a problem that often has less to do with the rat and more to do with man. Rats will always be the problem. Rats command a perverse celebrity status—nature's mobsters, flora and fauna's serial killers—because of their situation, because of their species-destroying habits, and because of their disease-carrying ability—especially their ability to carry the plague, which, during the Black Death of the Middle Ages, killed a third of the human population of Europe, something people remember, even though at the time people didn't know that rats had anything to do with all the panic, fear, and death.

In fact, in New York City, the bulk of rats live in quiet desperation, hiding beneath the table of man, under stress, skittering in fear, under siege by larger rats. Which brings me to my experiment: I went to the rat-filled alley to see the life of a rat in the city, to describe its habits and its habitat, to know a little about the place where it makes its home and its relationship to the very nearby people. To know the rat is to know its habitat, and to know the habitat of the rat is to know the city. I passed four seasons in the alley, though it was not a typical year by any definition. As it happened, shortly after I went downtown, the World Trade Center was destroyed. That fall, New York itself became an organism, an entity attacked and off-balance, a system of millions of people, many of whom were scared and panicked—a city that itself was trying to adapt, to stay alive. Eventually, New York regained its balance, and I went about my attempt to see the city from the point of view of its least revered inhabitants. And in the end—after seeing the refuse streams, the rat-infested dwellings, after learning about the old rat fights and learning all that I could learn from rat exterminators and after briefly traveling off from my alley to hear about rats all over America—I believe this is what I saw.

FOR MOST OF MY LIFE, however, my interest in rats had remained relatively idle, until the day I stumbled on a painting of rats by one of the patron saints of American naturalists, John James Audubon. Audubon famously documented the birds of North America in their natural habitat—*drawn from nature* was his trademark—and he next did the same for mammals, even the rat, or in this case several rats in a barn, stealing a chicken's egg. As I investigated the painting, I learned that Audubon had researched rats for months, and that in 1839 in New York City, where he lived during the last years of his life, he hunted rats along the waterfront. (He wrote the mayor and received permission "to shoot Rats at the Battery early in the morning, so as not to expose the inhabitants in the vicinity to danger . . .") In other words, Audubon was not just a Representative Man out of the American past whose legacy inspired American conservationists and environmentalists, not just some Emersonian model, but also a guy who spent time in New York City walking around downtown looking for rats.

I read more about Audubon. I read that he was born in what is now the Dominican Republic. I read that he turned to painting late in life after failing as a businessman, and that after traveling all over the continent to finish *The Birds of North America* he moved to New York, living first downtown, then up on what is today 157th Street, in a neighborhood that is coincidentally now settled by people from the Dominican Republic—coincidence is the stuff of ratting! I read that he fished in the Hudson River. I read that his eyesight eventually went, that shortly thereafter he began singing a French children's song over and over and eventually died. His home was left to rot away and was finally paved over. The more I read of Audubon, the more I felt a desire to study the rat in its urban habitat, to *draw the rat in nature.*

One day, I got on the subway and took a trip uptown. I went to Trinity Cemetery on 155th Street and saw the tall, animal-covered Celtic cross on Audubon's grave, and then, with old maps, I tried to figure out where his house would have been. Finally, I found the lot, unmarked; it had apparently once been on a gentle hill sloping toward the river, but now it was a hole, a three-story-deep pit, surrounded by two tall apartment buildings, and an elevated highway. When I looked away from the hole, the view was breathtakingly panoramic and Hudson River-filled. And when I got my binoculars out and looked down into the site, I could see the dozens of tennis-ball-size burrows that are more commonly referred to as rat holes.

*Chapter 2*

# THE CITY RAT

BUT ENOUGH ABOUT *you,* I think I hear the reader protesting. What about rats? And so, as I arise from my selfishness to describe the wild rat of New York City, the object of this nature experiment, I begin by noting that when it comes to rats, men and women labor under a lot of misinformation—errors inspired, it seems to me, by their own fears, by their own mental rat profiles rather than any earth-based facts. So, with this in mind, I offer a brief introductory sketch of the particular species of rat that runs wild in New York—*Rattus norvegicus,* aka the Norway or brown rat. I offer a portrait that is hysteria-free, that merely describes the rat as a rat.

A rat is a rodent, the most common mammal in the world. *Rattus norvegicus* is one of the approximately four hundred different kinds of rodents, and it is known by many names, each of which describes a trait or a perceived trait or sometimes a habitat: the earth rat, the roving rat, the barn rat, the field rat, the migratory rat, the house rat, the sewer rat, the water rat, the wharf rat, the alley rat, the gray rat, the brown rat, and the common rat. The average brown rat is large and stocky; it grows to be approximately sixteen inches long from its nose to its tail—the size of a large adult human male's foot—and weighs about a pound, though brown rats have been measured by scientists and exterminators at twenty inches and up to two pounds. The brown rat is sometimes confused with the black rat, or *Rattus rattus,* which is smaller and once inhabited New York City and all of the cities of America but, since *Rattus norvegicus* pushed it out, is now relegated to a minor role. (The two species still survive alongside each other in some Southern coastal cities and on the West Coast, in places like Los Angeles, for example, where the black rat lives in attics and palm trees.) The black rat is always a very dark gray, almost black, and the brown rat is gray or brown, with a belly that can be light gray, yellow, or even a pure-seeming white. One spring, beneath the Brooklyn Bridge, I saw a red-haired brown rat that had been run over by a car. Both pet rats and laboratory rats are *Rattus norvegicus,* but they are not wild and therefore, I would emphasize, not the subject of this book. Sometimes pet rats are called fancy rats. But if anyone has picked up this book to learn about fancy rats, then they should put this book down right away; none of the rats mentioned herein are at all fancy.*

Rats are nocturnal, and out in the night the brown rat's eyes are small and black and shiny; when a flashlight shines into them in the dark, the eyes of a rat light up like the eyes of a deer. Though it forages in darkness, the brown rat has poor eyesight. It makes up for this with, first of all, an excellent sense of smell. Rats often bite young children and infants on the face because of the smell of food residues on the children. (Many of the approximately 50,000 people bitten by rats every year are children.) They have an excellent sense of taste, detecting the most minute amounts of poison, down to one part per million. A brown rat has strong feet, the two front paws each equipped with four clawlike nails, the rear paws even longer and stronger. It can run and climb with

---

* Fancy rats are related to the wild *Rattus norvegicus* very possibly because of Jack Black, the rat catcher to Queen Victoria. Jack Black caught rats for the queen, but he also kept rats that interested him for himself. He sold some of these rats to women; in the Victorian era, keeping rats as pets was a fad—Beatrix Potter is thought to have bought her pet rat from Jack Black himself. Jack Black also bred a strain of albino *Rattus norvegicus* that he subsequently sold to Victorian-era scientists in France. Laboratory rats are today available for purchase on-line; a scientist can order the rat as per his or her experimental rat-genetics needs. The progenitor of the modern laboratory rat is the Wistar rat, a rat bred in the Wistar laboratories in Philadelphia. I have read that the Wistar rat was begun with an albino rat that the Wistar Institute originally got from France. I like to think that all the great scientific achievements that have been made in the modern scientific era as a result of work with laboratory rats are ultimately the result of the work of Jack Black, rat catcher.

squirrel-like agility. It is an excellent swimmer, surviving in rivers and bays, in sewer streams and toilet bowls.

The brown rat's teeth are yellow, the front two incisors being especially long and sharp, like buckteeth. When the brown rat bites, its front two teeth spread apart. When it gnaws, a flap of skin plugs the space behind its incisors. Hence, when the rat gnaws on indigestible materials—concrete or steel, for example—the shavings don't go down the rat's throat and kill it. Its incisors grow at a rate of five inches per year. Rats always gnaw, and no one is certain why—there are few modern rat studies. It is sometimes erroneously stated that the rat gnaws solely to limit the length of its incisors, which would otherwise grow out of its head, but this is not the case: the incisors wear down naturally. In terms of hardness, the brown rat's teeth are stronger than aluminum, copper, lead, and iron. They are comparable to steel. With the alligator-like structure of their jaws, rats can exert a biting pressure of up to seven thousand pounds per square inch. Rats, like mice, seem to be attracted to wires—to utility wires, computer wires, wires in vehicles, in addition to gas and water pipes. One rat expert theorizes that wires may be attractive to rats because of their resemblance to vines and the stalks of plants; cables are the vines of the city. By one estimate, 26 percent of all electric-cable breaks and 18 percent of all phone-cable disruptions are caused by rats. According to one study, as many as 25 percent of all fires of unknown origin are rat-caused. Rats chew electrical cables. Sitting in a nest of tattered rags and newspapers, in the floorboards of an old tenement, a rat gnaws the head of a match—the lightning in the city forest.

When it is not gnawing or feeding on trash, the brown rat digs. Anywhere there is dirt in a city, brown rats are likely to be digging—in parks, in flowerbeds, in little dirt-poor backyards. They dig holes to enter buildings and to make nests. Rat nests can be in the floorboards of apartments, in the waste-stuffed corners of subway stations, in sewers, or beneath old furniture in basements. "Cluttered and unkempt alleyways in cities provide ideal rat habitat, especially those alleyways associated with food-serving establishments," writes Robert Corrigan in *Rodent Control*, a pest control manual. "Alley rats can forage safely within the shadows created by the alleyway, as well as quickly retreat to the safety of cover in these narrow channels." Often, rats burrow under concrete sidewalk slabs. Entrance to a typical under-the-sidewalk rat's nest is gained through a two-inch-wide hole—their skeletons collapse and they can squeeze into a hole as small as three quarters of an inch wide, the average width of their skull. This tunnel then travels about a foot down to where it widens into a nest or den. The den is lined with soft debris, often shredded plastic garbage or shopping bags, but sometimes even grasses or plants; some rat nests have been found stuffed with the gnawed shavings of the wood-based, spring-loaded snap traps that are used in attempts to kill them. The back of the den then narrows into a long tunnel that opens up on another hole back on the street. This second hole is called a bolt hole; it is an emergency exit. A bolt hole is typically covered lightly with dirt or trash—camouflage. Sometimes there are networks of burrows, which can stretch beneath a few concrete squares on a sidewalk, or a number of backyards, or even an entire city block—when *Rattus norvegicus* first came to Selkirk, England, in 1776, there were so many burrows that people feared the town might sink. Rats can also nest in basements, sewers, manholes, abandoned pipes of any kind, floorboards, or any hole or depression. "Often," Robert Corrigan writes, " 'city rats' will live unbeknownst to people right beneath their feet."

Rats also inhabit subways, as most people in New York City and any city with a subway system are well aware. Every once in a while, there are reports of rats boarding trains, but for the most part rats stay on the tracks—subway workers I have talked to refer to rats as "track rabbits." People tend to think that the subways are filled with rats, but in fact rats are not everywhere in the system; they live in the subways according to the supply of discarded human food and sewer leaks. Sometimes, rats use the subway purely for nesting purposes; they find ways through the walls of the subway stations leading from the tracks to the restaurants and stores on the street—the vibrations of subway trains tend to create rat-size cracks and holes. Many subway rats tend to live near stations that are themselves near fast-food restaurants. At the various subway stations near Herald Square, for example, people come down from the streets and throw the food that they have not eaten onto the tracks, along with newspapers and soda bottles and, I have noticed, thousands of no-longer-charged AA batteries, waiting to leak acid. The rats eat freely from the waste and sit at the side of the little streams of creamy brown sewery water that flows between the rails. They sip the water the way rats do, either with their front paws or by scooping it up with their incisors.

DEATH COMES IN MANY FORMS for a brown rat living in the wilds of the city. A rat can be run over by a car or a bus or a cab. It can be beaten with a plunger as it climbs up through a sewer pipe and surfaces into an apartment's toilet bowl. Cats, while mice eaters, are not likely to attack adult rats; a rat will easily repel an attack by a cat, though cats will kill young rats. In the city's less populated areas, or in the little patches of parkland and green, rats sometimes die quasi-wilderness deaths. In Brooklyn's Prospect Park, I once watched a large red-tailed hawk swoop down on a brown rat, an adult male that had been living in a burrow in a wooded area adjacent to an overstuffed garbage can. The hawk then flew into the upper branches of a maple tree, dangling the large, still-wriggling rat from its talons. People have confided rat shootings to me on numerous occasions; in fact, more people than I had ever imagined shoot rats in the city—using pellet guns or air rifles or even more potent rifles in alleys and in infested basements. And of course, rats also die when they are caught in snap traps, which is the trap sometimes referred to as a break-back trap, a rat-size version of the classic mousetrap. It is especially difficult to trap a rat with a snap trap. Generally speaking, rodents are wary of new things in their habitat, preferring routine to change; biologists refer to this trait as neophobia. Rats can be even more neophobic than mice. Thus, exterminators are likely to leave unset snap traps out for a few days before setting them, often baited, allowing the rats to become comfortable with traps. Some exterminators regularly treat snap traps with bacon grease.

Most frequently rats die from ingesting poison. I don't know of a precise statistic, but I know that at any given moment there is poison all over the streets and homes of New York, not to mention the rest of

America. Sometimes, poison is injected directly into the rat burrow; the rat dies of heart failure or, with the most severe poisons, of damage to its central nervous system—they are found dead on their bellies, arms and legs extended. More often, poison is added to grains and the grain is put into shoe-box-size containers called bait stations. Bait stations are the things that people in cities see constantly in back alleys and in parks and do not recognize or, chances are, even think about. Bait stations are designed to keep bait away from pets and children, but they are also designed as little rat-friendly zones. With their small holes and zigzaggy interiors, bait stations are to a rat what a smoothly run fast-food restaurant is to a human. When rats eat the poisoned grain in the bait station, they return to their nests to die—in walls, in floors, underneath streets and restaurant stoves, in sewers. The most widely used poisons are anticoagulants, which cause the rat to bleed to death internally. It takes several meals for the rat to die. As it returns, it sometimes seems more and more woozy. Exterminators refer to this phenomenon as "dead rodent walking."

Ingesting poison, fighting for food, being attacked by a larger rat or beaten with a toilet plunger: these are everyday rat dangers that make the life expectancy of the rat in the city approximately one year. And yet rats persist; they thrive in New York City and in cities throughout the world. Rats do not inhabit cities exclusively, of course; like man, rats can live anywhere. Brown rats in wilderness areas are sometimes called feral rats; they survive on plants and insects and even swim to catch fish.* However, brown rats are generally larger and more

* One of the most impressive examples of a rat incursion in a noncity area was the invasion of brown rats on Campbell Island, a remote patch of land south of New Zealand near Antarctica. They were thought to have been imported to the island by whaling ships in the nineteenth century. The rats destroyed the local population of birds, including a rare flightless teal and a wading duck. In 2002, the New Zealand government destroyed all of the rats by bringing 120 tons of rat poison to the island in boats and helicopters. Approximately 200,000 rats are thought to have died. The rat eradication, frequently referred to as the largest-ever rat hunt, encountered some problems, however. A tanker carrying 18 tons of rat poison to the island sank in a whale breeding ground. The rat poison has subsequently showed up in the local mussel population. Reports suggested that all of the rats were killed. The government hopes to reintroduce the teal and the duck.

numerous in cities. As a result, it is in cities that they are especially successful at spreading the diseases that are like poisons to humans. They carry diseases that we know of and they may carry diseases that we do not know of—in just the past century, rats have been responsible for the death of more than ten million people. Rats carry bacteria, viruses, protozoa, and fungi; they carry mites, fleas, lice, and ticks; rats spread trichinosis, tularemia, leptospirosis. They carry microbes up from the underground streams of sewage; public health specialists sometimes refer to rats as "germ elevators." Though targeted over and over by man, rats generally wreak havoc on food supplies, destroying or contaminating crops and stored foods everywhere. Some estimates suggest that as much as one third of the world's food supply is destroyed by rats.

Rats succeed while under constant siege because they have an astounding rate of reproduction. If they are not eating, then rats are usually having sex. Most likely, if you are in New York while you are reading this sentence or even in any other major city in America, then you are in proximity to two or more rats having sex. Male and female rats may have sex twenty times a day, and a male rat will have sex with as many female rats as possible—according to one report, a dominant male rat may mate with up to twenty female rats in just six hours. (Male rats exiled from their nest by more aggressive male rats will also live in all-male rat colonies and have sex with the other male rats.) The gestation period for a pregnant female rat is twenty-one days, the average litter between eight to ten pups. And a female rat can become pregnant immediately after giving birth. If there is a healthy amount of garbage for the rats to eat, then a female rat will produce up to twelve litters of twenty rats each a year. One rat's nest can turn into a rat colony of fifty rats in six months. One pair of rats has the potential of 15,000 descendants in a year. This is a lot of rats, and while the regenerative capabilities of the rat might seem incomparable to those of any other species, in *Rats, Lice, and History*, the classic work on the effect of disease on human history, Hans Zinsser suggests that the fertility rate of the human can rival the fertility rate of the rat.

★ ★ ★

One of the things I find most fascinating about rats is that they have a sense of where they are and of where they have been. This is explained by the fact that rats love to be touching things. Biologists refer to rats as thigmophilic, which means *touch loving*. Consequently, rats prefer to touch things as they travel. Their runways are often parallel to walls, tracks, and curbs; in infested basements, grease slicks parallel ceiling beams and the run of sewer pipes. Rats are thought to feel especially safe at corners, when they are simultaneously touching a wall and free to escape. As they travel again and again for food, as they escape oncoming trucks or, upon the return home of a drunk human apartment dweller, flee into the relative safety of garbage cans, rats develop a muscle memory, a kinesthetic sense that allows them to remember the turns, the route, the course of movement. As young rats follow older rats, the trails are repeated, passed on. Exterminators like to say that if the walls of an alley or a rat-infested block were somehow taken down without disturbing the rats, the rats would awaken the next evening, venture forth, and travel precisely the same routes as the night before, as if the walls were still there. They would remember the walls. Deep in their rat tendons, rats know history.

A rat phenomenon that is based only partly on fact is the Rat King, a kind of rat often mentioned in stories about rats. The Rat King is usually described as the rat that leads other rats when rats amass and herd. Policemen on late night patrols sometimes report seeing a Rat King lead a group of rats across a street. Drunks frequently report Rat King sightings. It *is* true that from time to time rats run in huge packs. I have seen them do so. Likewise, it is true that within a rat colony a dominant male rat emerges. However, it is not the case that one rat leads the others. Something that has inspired the notion of a mythical Rat King is the actual phenomenon of rats whose tails have become knotted together with other rats' tails in their nest. The resulting entanglement is called a Rat King. There have been Rat Kings ranging in size from three rats to thirty-two rats. Sometimes the rats die, sometimes they are fed by the other rats and stay alive for a time in the nest. In myths and stories about marauding rats and secret rat leagues, the Rat King sometimes sits in the center of tied-up rats' tails, the lesser rats his throne. But again, these are rat stories. An actual Rat King is really nothing more than a rat that takes advantage of his natural strengths and of other rats' natural weaknesses. A Rat King is just a big rat.

THE RAT IS A NEWCOMER to America, an immigrant, a settler, its ancient roots reaching to Southeast Asia. The black rat migrated south, while the brown rat migrated north, to China, along the Yangtze River, and then into Siberia near the present-day Lake Baikal. The black rat came to Europe ahead of the brown rat, with the Crusades. The brown rat did not appear in Europe until the beginning of the eighteenth century. There are accounts of brown rats crossing the Volga River in hordes in 1727, and more reports of brown rats proceeding across Russia to the Baltic Sea. Brown rats were reported in east Prussia, France, and Italy in 1750; they were reported in Norway in 1768 and in Sweden in 1790. Brown rats are thought to have been brought by ship from Russia to Copenhagen in 1716 and to Norway from Russia in 1768. Spain did not have brown rats until 1800. They arrived in England in 1728, and in 1769, in *Outlines of the Natural History of Great Britain*, John Berkenhout named the brown rat *Rattus norvegicus*. He most likely misnamed the rat. He believed that the rats had come to England via Norwegian lumber ships, when in fact they had probably come from Denmark, since at the time Norway rats had not yet settled in Norway.

By 1926, *Rattus norvegicus* was in every state in America. It pushed out black rats everywhere, though a small colony of black rats held on in New England for many years. The last state to be settled by *Rattus norvegicus* was Montana. Several early rat settlements in Montana failed or were wiped out with poisons and traps, but the brown rat finally colonized Lewistown in 1920, and in 1938, the dump in Missoula was the site of an escaped colony of laboratory rats, domesticated *Rattus norvegicus*. It was not easy for the brown rat to settle Montana. "In general it appears that rats find extension in their range difficult in

Montana and that in all likelihood this difficulty is due to the sparseness of the population," a biologist in Bozeman wrote. Brown rats also eventually spread to all the provinces of Canada, with the exception of Alberta, where in 1950 they were reported on the southeast border but were then repelled by an intensive government rat control program, one of the most impressive rat-control programs in the world. Alberta still considers itself, in the words of the province's agricultural department, "an essentially rat-free province."

Little is written about the early settlement of *Rattus norvegicus* in America. Most reports state that the very first *Rattus norvegicus* arrived in America in the first year of the Revolution, then moved out into the country, a manifest infestation. One of their first landings was most likely New York City.

## *Chapter 3*

# WHERE I WENT TO SEE RATS AND WHO SENT ME THERE

IN GOING TO my alley, I was going where someone had gone before, of course, and I'm not just thinking of the millions of people who walk by it every year or the inebriated souls who stumble into it accidentally or the people who step into it because they think it is an actual street, which it isn't. I'm thinking of David E. Davis, the founding father of modern rat studies. It is said that Alexander the Great kept a copy of the *Iliad* in a precious casket as he went into battle, and in the same way I kept the work of Dave Davis beside me as I sat in the alley and excitedly took notes: the little diagrams that show rats running in dilapidated tenement neighborhoods, from fetid outhouse to poorly maintained garbage area; the field observations that look like maps with idle doodlings. It was Davis who first documented the habits of the rat, who first charted their moves, who applied to a rat in an alley in a city the same kind of close nature reporting used, for instance, on the threatened marbled murrelet in its habitat in the Northwest coastal forest.

Davis began studying rats during World War II. The U.S. government was concerned that the Germans might use rats to spread disease through Europe, and then, after the war, with Europe's infrastructure in ruins, the government was concerned about rats ruining food supplies and spreading disease on their own. The Rodent Ecology Project was founded in Baltimore at Johns Hopkins University, and Davis worked there with the other founding fathers of rat studies: Robert Emlen, who, prior to working with Dave Davis, also worked with Aldo Leopold, an ecologist in Wisconsin who argued for a "land ethic," suggesting that humans ought to think about the relationship with the land on which they live; John Calhoun, who studied rat social behavior and in 1963 reported that rats left to overpopulate in a cramped room set about killing, sexually assaulting, and cannibalizing each other; Curt Richter, who started trapping rats with Davis, then went on to discover similarities between rats' and humans' diets and began experimenting on rats in laboratories, which led to all kinds of man-related studies on laboratory rats, like the one I read about in a newspaper recently that showed how rats will kill themselves overexercising; and finally, William Jackson, who advised governments around the world on rat control and rat poisons and then on what to do about rats that became immune to rat poisons. The scientists in the Rodent Ecology Project were working at the dawn of ecology, studying the relationship of an organism to its environment and to its fellow organisms and doing so with an organism that, frankly, no one wanted to have any kind of relationship with. They were World War II-era scientists who looked like World War II-era scientists—in photographs they wear short-sleeve, button-down shirts, khakis, pens in pockets. They went into neighborhoods that didn't see a lot of scientists—beat-up, run-down, near-the-waterfront neighborhoods, neighborhoods filled with people living in old tenement buildings, with people who due to poverty lived alongside rats. It was a new frontier for wildlife biologists. As P. Quentin Tomich, a biologist who worked with Davis as a graduate student and subsequently went off to study plague in rodents in Hawaii, told me, "No one had thought of the urban slums as a habitat."

Davis trapped rats, marked them, released them, trapped them again, and his papers opened a floodgate of myth-busting and groundbreaking rat information. "Although the brown rat *(Rattus norvegicus)* is an ubiquitous pest throughout the world," Davis wrote, "few studies of its home range and movements have been conducted." Davis showed that rats, commonly thought to be wanderers, in fact live in small areas, in colonies; that rats generally stay within sixty-five feet of their nest; that rats, when released far from their nest, will nonetheless wander for miles (up to four miles in one study); that male rats tend to go farther away from their nest than female rats; that one way rats may protect themselves is by becoming completely familiar with their home territory, their city alley or block ("Thus an individual that knows every hole, bush, or shelter probably will escape enemies better than an individual unacquainted with the area," Davis wrote); that rats are likely to cross alleys but not roads; that rats use regular runways or paths to feed, taking the same paths night after night, rarely diverging, rarely straying ("For example, a rat may live under the steps, run along the fence to the alley, and there feed on garbage," he said); that rats in the city are often bigger than rats in the country; and that the social rankings of the rat colony are of great significance, especially in times of duress—the strong rats thrive, while the weaker rats begin to die. In "Characteristics of the Global Rat Populations," an article published in the *American Journal of Health* in 1951, Davis wrote, "As the population increases relative to its food supply, the higher ranking members still get adequate food, but the low members begin to starve. Low-ranking females have poor reproductive success and progeny from low-ranking females have little chance to grow normally."

Throughout the 1950s, Davis was America's rodent control guru. He traveled America's rat populations. He consulted with cities on their rats, preaching his most important discovery throughout the country—that poisoning rats was not in itself an effective way of controlling them. In fact, when rats are killed off, the pregnancy rates of the surviving rats double and the survivors rapidly gain weight. The rats that survive become stronger. "Actually, the removal merely made room for more rats," Davis wrote. The only way to get rid of rats was to get rid of the rat food, or garbage, but no one wanted to hear this: as it was the dawn of the age of ecology so also it was the dawn of the age of the chemical, of poisons and pesticides, and people seemed to want a sexier, chemical-based fix. Eventually, Davis became frustrated. He moved to Pennsylvania, where he studied animals other than rats. For a while, he studied woodchucks and once sent a colony of them on a boat to Australia in a darkened box to see how the trip to the other side of the world would affect their internal clock: on the ship, they stayed on Pennsylvania time, but when the box was opened in Australia, they switched immediately to Australian time. He arranged a grid over a field to study birds. His three daughters remember him waking up at 5 A.M., sitting in his bathrobe, looking into the backyard sky, speaking into a tape recorder and saying things like "Three starling flying away from the city."

Davis taught in North Carolina and then California, and in his retirement, he wrote a paper that applied his many years of animal ecology and population studies to human history; it was published by his daughters in 1995, a year after his death. In the paper, he posited that the great cathedrals of Europe were a result of an excess food supply for the human population at the time. To read this paper is to see that thinking about rats, as low-down as it seems, can easily lead to thoughts about larger topics, such as life and death and the nature of man. "The population trebled in three centuries," Davis wrote. "As the population reached the capacity level of food and other resources, its growth stopped, and construction of cathedrals ended. The period terminated in wars, litigation, and disease. The hypothesis arose from the study of principles of population as derived from experiments on animals such as mice. Obviously, the test has not been experimental. A test that has many elements of an experiment is now possible in the oil-producing nations. These countries have suddenly found a source of energy. They will develop new types of art, literature, and science and will build vast structures not yet conceived. Then, as the population reaches the limit of resources (a complex stage involving the entire world), the period of history will end in stagnation, conflict, and misery. Humans have the knowledge to prevent a repetition of the later history of the Middle Ages."

A forgotten accomplishment of Dave Davis is his debunking of what is still today the most often quoted statistic about rats—the one rat per person rule. This statistic is ritually used in news stories about rats and has been for almost a hundred years. It is not true. It is a bastardization of a statistic derived from a study of rats written in England in 1909 by W. R. Boelter, entitled *The Rat Problem.* At the time, Boelter toured the English countryside and asked the following question: "Is it reasonable to assume that there is one rat per acre?" People responded by saying things such as "certainly" or "absurdly low." Boelter did not ask people in cities the same question because he thought it was ridiculous to ask people in the city if they had a rat. "As regards villages, towns, and cities, I consider it unnecessary to ask a question, the answers to which must be obvious to anyone who thinks of the number of pantries, houses, shops, stores, and sewers to be found on one acre," Boelter wrote. In the end, he made an educated guess: one rat per acre in England. And because there were forty million cultivated acres in England at the time, he concluded that there were forty million rats. *Coincidentally,* forty million people lived in England in 1909. Boelter was able to convert the one-rat-per-acre statistic to one rat per human. People loved that statistic, maybe because they abhorred it. They did not bother to recalculate for their own particular rat and human populations—an extremely labor-intensive process that at the time only Davis seemed to be interested in executing. Subsequently, one-rat-per-human has become the sacred rat statistic. The United Nations has used it. Pest control companies use it; health departments use it. Even today, it is commonly said that in New York there are eight million rats, one for every New Yorker.

In 1949, Dave Davis analyzed New York's rat population and called the one-rat-per-human statistic "absurd." He had just completed a precise calculation of the rat population of Baltimore—by trapping, counting burrows, and measuring such things as rat runways and rat droppings. In New York, he began his work on six blocks in East Harlem. He brought in an experienced trapper to trap rats in East Harlem apartments for a week. Davis determined there were an average of three rats per apartment in infested Harlem buildings, mostly living in the kitchen and bathroom but traveling through many floors. He further determined that more people thought they had rats than actually had them—about 10 percent more. But when he added up his calculations, New York's rat population was nowhere near eight million. Even the New York waterfront, which was mythically associated with rats, was less infested than assumed. "Certainly, there are no more than a few thousand in the entire dock areas of New York City," Davis wrote. In all, Davis put the rat population of New York at one rat for every thirty-six people, or 250,000 rats—a rat population the size of the human population of Akron, Ohio. When the health department read Davis's report, they canceled a citywide rat extermination plan. But the number-of-humans-equals-number-of-rats formula would not die. It is something people want to believe. A few years later, even the New York City health department was telling people that there were eight million rats in New York.

WHEN I FINALLY WENT OUT on my own to find a colony of wild New York City rats, I ended up talking to a lot of exterminators. Exterminators, or pest control technicians as they often prefer to be known, are the philosopher kings of the rat-infested world, the trap- and poison-toting mystics. I have gleaned many insights from them. Practically speaking, I have learned about the significance of spotting rats during the day. "When you see rats in the daytime, boy, the population is so large that the night feeding won't support them," one exterminator told me. "Only the dominant rats are getting enough to eat, and the weaker rats, they've gotta take a chance and go out during the day. They don't really want to be out during the day." Likewise, I learned about the strength of rats vis-à-vis cats. Here is this anecdote from an exterminator working in New York, in the borough of Queens: "A woman said to me, 'Oh, we're going to get a cat!' " he recalled. "I said, 'Miss, please don't put that cat in the cellar.' Then I came back two weeks later and I'm picking up the hair and the bones of the cat. They think it's like in the cartoons. But in the cartoons it's Tom and Jerry the *mouse,* not Tom and Jerry the *rat!*"

More than anything, I have learned from exterminators that history is crucial in effective rat analysis. In fact, history is everything when it comes to looking at rats—though it is not the history that you generally read; it is the unwritten history. Rats wind up in the disused vaults, in long underground tunnels that aren't necessarily going anywhere; they wind up in places that are neglected and overlooked, places with a story that has been forgotten for one reason or another. And to find a rat, a lot of times you have to look at what a place was. One exterminator I know tells the story of a job on the Lower East Side in an old building where rats kept appearing, nesting, multiplying, no matter how many were killed. The exterminator searched and searched. At last, he found an old tunnel covered by floorboards, a passageway that headed toward the East River. The tunnel was full of rats. Later, he discovered that the building had housed a speakeasy during Prohibition. After figuring out a place, after getting to know it intimately, killing rats is the easy part. "The textbook scenario, if you want to get rid of rats, is you put stress on their environment, you stop the food, and then they eat each other," another exterminator told me.

In beginning my own search for a rat colony, I turned to one exterminator frequently, George Ladd. George is tall with short, bristly hair; he is in his mid-fifties and fit and he routinely wears a blazer and tie while out on his pest control jobs, looking less like someone who hunts vermin for a living and more like a college coach dressed up for a big game. He works out of an office on the Lower East Side that from the outside looks like it's going to be a real mess but ends up being immaculate. And like many exterminators, he not only knows a lot about rats, but also about how humans relate to them. "You get a call and you just know right away from the intensity of the person calling whether or not they've got rats," he says. George has a lot of respect for rats. "They're just rude. They're like a bear because they're smart. They're *extremely* smart animals."

Ladd rides from job to job on a motorcycle, and he calls his company Bonzai de Bug. The Japanese reference has to do with George Ladd's great-grandfather George Trumball Ladd. George Trumball Ladd was a philosopher and a founder of the modern science of psychology; along with his friend William James, he organized the American Psychological Association. George's grandfather was also an influential figure in Japan; he was a friend and adviser to the emperor, promoting friendly relations between Japan and America, until he died in 1921. Half of his ashes are buried in Japan, half in New Haven, Connecticut, where he taught at Yale. In Japan, George Ladd is a minor celebrity.

Once, when I dropped in on George, he showed me a videotape of a program about him that had run on Japanese TV. He put the tape in and started it up, but then stopped it to take a call from a landlord dealing with a tenant who had not realized that an exterminator was coming by and either didn't hear the buzzer or chose not to get out of bed and answer the door and, thus, was startled to see an exterminator—one of George's assistants—in her room. When George hung up, he started the tape again. The program began with ominous music and images of New York City's skyline, then New York City's trash. Next was a shot of George riding into town on his motorcycle and buckling on a belt full of rat-fighting equipment. The show was in Japanese but George spoke in English.

"How fun is this?" George said, smiling and pointing to himself as we watched.

The Japanese program showed George talking to various people plagued by rats, and it showed him working in a building late one evening. As I watched, George recalled the particular rat problem that the TV show was documenting; it involved an apartment building in a fashionable area, the kind of neighborhood where people instruct the exterminators to work in secret, so that no one will know they have rats. "They had a rat in a fancy-ass building on the Upper West Side, and we couldn't even *think* about traps or bait or anything," George said. "We had to get 'em, period. So I went to the store and I bought Hershey's bars, nuts—they love nuts—anchovies, beer. They drink beer and they like it, but they drink a lot and then they can't throw up. And then shrimp. Then I rubbed the shrimp around the edges. Then I took it and I put it in the center. It had a nice strong smell."

I asked him what happened next.

"I nailed that thing on the first night," he said.

On TV, the Japanese film crew showed him going into a bodega, buying the food items that he had just mentioned, and pasting them down to a big glue-covered board. As he affixed the food items to the glue, George said, "Soup's on!"

In the next scene, he returned to the building the following day to inspect a dead, glued-to-the-board rat. "Got that sucker," George said.

On the TV show, he looked into the camera and spoke: "Being calm, cool, collected—it's all part of my job."

Eventually, I asked George where he thought I should go to study rats. He suggested the Seaport District, a place I was especially interested in. "Check out down around Gold Street," he said. "I've seen some big ones down there."

SHORTLY THEREAFTER, IN THE SPRING of 2001, I was ready to begin a year of rat watching, four seasons spent among vermin. On the afternoon of the night I first went out, I purchased a night-vision monocular. "This is the model that everybody looking for rats uses," the salesman said to me, facetiously, I think. And then, on my way home, when I spotted a rat on the subway tracks, I found myself taking it out to use in the station. I ran alongside the rat on the platform, until I realized everyone in the station was watching me, at which point I put the night-vision monocular away and tried to play it cool. That night, I was out in the warm spring darkness, out in lower Manhattan, out in the nearly deserted, late-night city—the habitat of the *Rattus norvegicus*. I was down near the bottom of Broadway, in the oldest quarter of Manhattan, the place where the city began. I passed City Hall, lights on its neoclassical front, noisy starlings in the London plane trees. I was out ratting.

I turned down Beekman Street, and in a few feet I was looking into Theatre Alley. If ever an alley looked like a rat-infested alley, then Theatre Alley is it: steam rising from a narrow, old cobblestone street like fog in a Hollywood horror film, high walls of trash that appeared to be formed by rock slides—a lost side canyon. The little bit of exterminator-inspired historical research that I'd done told me that Theatre Alley has always been a place for rats. In the nineteenth century, there was a grog shop on the alley, which was swill-filled and pig-roamed, littered sometimes with the corpses of dead horses, like many streets of New York at the time. An old British theater backed onto the alley, the theater that inspired the British spelling of Theatre Alley's name. And—because all the newsrooms of the city's newspapers surrounded City Hall, because the neighborhood was littered with theaters and fancy hotels—the alley was the nineteenth-century equivalent of a shortcut through Times Square, a timesaving runway for the entertainment-seeking masses, for Knickerbocker New York's publishing crowd.

In my mind's eye, as I looked down Theatre Alley, I could see the crowds that had snuck through it, and I even like to think I could see individuals. There was Herman Melville, who used a library at one end of the alley. There was Edgar Allan Poe, who edited a newspaper a block away; I could see Poe trudge through garbage, see rats scatter. There was Walt Whitman, who worked at a newspaper just around the corner and loved to go to plays, who might have walked up Broadway the time he heard Emerson speak, and then, when he typed up his review of Emerson's lecture for his newspaper, might have walked back through Theatre Alley thinking about the soul and nature as an expression of God—or so I like to think. "I embrace the common, I explore and sit at the feet of the familiar, the low," Emerson once wrote.

Unfortunately, I couldn't see any rats.

My luck changed around the corner. On Nassau Street, I found an abandoned fast-food place, its back to Theatre Alley. I looked in the window and at first saw just a flash of movement in the back corner, where long ago there had maybe been a shake machine or an area for warming fries; I noticed it the way a birder might notice the flush of the bird's tail. I looked more closely, using my binoculars—night-vision gear was not necessary given that there was some light in the back of the restaurant—until I saw it: the dark gray, bricklike chunkiness of it—a rat climbing out of a hole in the wall and hitting the floor. More rats arrived, scurrying around, just as fast-food patrons might previously have done, all coming from the lot next door, a mostly broken-brick-filled hole, the remains of a recently demolished building that had once stood on Theatre Alley. And yet, the place was not ideal for my purposes. The rats were behind fast-food restaurant glass. They were relatively far away. It was a good place to see rats from afar, but not a good place to study them up close.

I pressed on, continuing down Fulton Street, walking toward the waterfront through the trickle of tired tourists and late-night bargoers beginning to wander home. I searched for rats around Water Street and then near Peck Slip, a boat slip in the time of the Revolutionary War that was subsequently filled in, that was made dry land by dumping decrepit boats and garbage of all kinds, and is now a little run-down cobblestone piazza, framed by the nighttime lights of the Brooklyn Bridge. I could smell fish from the Fulton Fish Market, a rat-inspiring scent, and I had hoped there would be some garbage in the streets but it was relatively clean—the sanitation department's trucks had just been through. On Front Street, I used my night-vision gear to peer into lot after lot, but I saw nothing except four tall, thin, young women who walked past me in a little pack, their burning red-tipped cigarettes glowing before them.

Frustration was what I was feeling, to be sure, for it was almost midnight, and I was wondering if I would ever find a good rat colony—and yet it was exciting to be out when the city was pared down to its late-night self. With the sidewalks less peopled and the garbage bags hauled out, in the fluorescent-yellow laboratory glow of the streetlights, human activity seemed more noticeable, accentuated; nighttime is a great time to explore the brown rat's habitat. At the intersection of Peck Slip and Water Street, a large green garbage truck stopped behind a broken-down taxicab that was blocking the one-way street. Men piled off of the truck and began to push the cab out of the street—they rocked it and shoved it, while the cabdriver pushed and steered the car to the side of the road and shouted thank-yous. The men from the garbage truck cheered and embraced and shook hands, one by one, with the cabbie—and then watched in wonderment as a man in a business suit wobbled out of nowhere and, while yelling into a cell phone, tried again and again to hail the broken-down cab.

At last, a few blocks back toward land, I came to an alley that I'd never been down before, that I'd never even noticed. Sure enough, it was just off Gold Street, the area George Ladd had suggested. It was a shortcut from Fulton to Gold Street, connecting the two like an elbow, a quicker way by a few seconds from the Seaport to, say, Wall Street, a few blocks away—a little place called Edens Alley.

I walked through the alley that would become my base for four seasons and did not see rats at first—though the alley smelled of urine, and dark green garbage bags lined it like ornamental hedges. Upon reaching the corner of the alley, I looked into what seemed like an abandoned lot. There was rustling, the sound of something moving. I stood quietly, so that soon the alley in which nothing seemed to be happening was filled with movement. When I looked into the lot with my night-vision gear, I saw, first, little bright eyes, shining in the infrared gaze, and next, more little bright eyes.

There were rats in Edens Alley, all right, and there were lots of them, all scurrying around in the dark. Some of the rats were larger than the other rats, some of the rats were smaller, and they were all running around, carrying food, burrowing in a pile of what appeared to be sand and then disappearing and returning again in a way that, since I was not at all familiar with such a scene, kind of made my skin crawl. The rats were busy, alive. They were rat-happy rats. I immediately recognized this alley as an excellent natural wild-rat habitat.

*Chapter 4*

# EDENS ALLEY

TO KNOW THE city rat, as I have been saying, one has to know the rat's habitat, which is of course the city. And if the city is a world unto itself, then the inhabitant of the city knows only the littlest speck, or as Thoreau has written in *Walden*: "We are acquainted with a mere pellicle of the globe on which we live. Most have not delved six feet beneath the surface." Late that spring I delved into the pellicle that is Edens Alley and set about surveying the alley that was my city rat's home range. I was off into the alley itself by night, into old books and old city newspapers by day.

To begin with, Edens Alley is L-shaped, a cobblestone strip that is surrounded by brick walls—a walled-in lane that was like my Walden, though I'm not so nuts as to want to actually sleep there or anything. The lower portion of the L is narrow, only about twelve feet wide, and although it looks as if it were a dead end, it makes a ninety-degree left-hand turn to travel another two hundred feet until it comes back to a main thoroughfare. Technically speaking, the longer portion of the alley is called Ryders Alley; the shorter portion is Edens Alley. Altogether it is about as long as a suburban driveway, though there is nothing suburban about the alley. While a suburban yard is the bastardization of rural ideals, maintained with fertilizers and landscaping teams, this alley might be described as a symbol of the urban *un*-ideal, the not seemingly necessary evil: squalor and neglect, refuse and, of course, vermin. Also you could never grow a lawn in Edens Alley. Even during the day there is little sun.

The alley is a nowhere in the center of everything. It is equidistant to the water (New York Harbor) and the financial capital of the world (Wall Street) as well as the city's governmental headquarters (City Hall), each of which is just a few blocks away. The alley is five blocks to Broadway, and if you step outside of the alley and walk a little and stand in the right spot, you can see the Empire State Building about fifty-five blocks north; as I mentioned, when I began visiting it, the alley was a few blocks from the World Trade Center. If you emerge from the alley at rush hour or at lunchtime on a weekday, then you get knocked into the flow of people as they herd to and from their offices, from their jobs and homes, as they follow their daily paths. But even then, even at the height of alley-area traffic, you can crawl back in the alley itself and be removed from the crowd, as if it were a hole. You can look out on the people as they pass, feel a quietness, a fermata of urbanity. The alley is a little secret street in the core of workaday, moneymaking, lunch- and breakfast-eating America.

The scenery of Edens Alley is on the humblest of scales. A sign at the entrance to the alley on Fulton Street says NO THRU TRAFFIC, and both people and cars obey—in the hundreds of days that I have visited the rat alley, I've never seen a car or a truck drive though it, the exception being garbage and food-delivery trucks, the vehicles of civic digestion. In a year, I only saw one truck regularly parked there—an unmarked van, the wheels of which rats would sometimes use for cover. As far as foot traffic goes, the alley is rarely profaned by passersby. People either do not realize that the alley is a very slight shortcut from Gold Street to Fulton Street, and vice versa, or, in weighing the benefits of possible time saved, they choose the extra distance over a trudge through this dark, garbage-littered, urine-scented path.

At the entrance on Fulton Street, there is an old eleven-story building; a vitamin and health-supplement store is on the ground floor and a Chinese restaurant is above it. On the other side is an apartment building with a souvenir store on its ground floor. When I first began studying rats there, in the summer, the souvenir store sold numerous New York City souvenirs—e.g., New York T-shirts and coffee cups and shot glasses—in addition to model birds that flapped their mechanical wings like pigeons. Walking up the southern edge of

Ryders Alley along the narrow, often garbage-covered sidewalk, one passes the following: an empty brick wall, three dark windows covered with medieval-seeming grates, a back door to the Chinese restaurant, a garage that seems to be an entrance to a dry-cleaning place, and a dark hole accessed by an old steel fire escape's staircase—the stairs angle down into complete darkness. Walking along the northern edge, you pass a long brick wall, a back door to an Irish bar and restaurant, a boarded-up window, another door to I don't know what, and a walled-in lot that is used for storage of construction materials. This lot was the place where I first saw rats in the alley, at play in the lot's sand.

As far as living things go, if you look up over the lot, then you see the alley's one nonconcrete event, a tree. It is an ailanthus tree. Generally speaking, the ailanthus thrives in places where other trees do not find a way to live. Sometimes, it is called "the tree that grows in Brooklyn." Its scientific name is *Ailanthus altissima*—literally, "tallest of the trees of heaven"; its common name is the tree of heaven. The tree of heaven is not considered native to New York; it is originally from China.* But it

* The term *native* when used in regards to plants and animals can be complicated. In an essay entitled "The Mania for Native Plants in Nazi Germany," published in a collection called *Concrete Jungle*, Joachim Wolschke-Bulmahn, the director of Studies in Landscape Architecture at Dunbarton Oaks, in Washington, D.C., says, "The missionary zeal with which so-called foreign plants are condemned as aggressive is significant. Such characterizations do not contribute to a rational discussion about the future development of our natural and cultural environment, but possibly promote xenophobia." Wolschke-Bulmahn points out that some plants that are considered "native" to the United States may have been carried over from Siberia by people migrating to America over a land bridge, and he writes of an early proponent of native plants, Jens Jensen, a landscape architect who lived in Wisconsin, who advocated the destruction of "foreign" plants, especially "Latin" or "Oriental" plants. Jensen had close ties to Nazi landscape architects in Germany. In a journal, Jensen wrote: "The gardens that I created myself. . . shall express a spirit of America, and therefore shall be free of foreign character as far as possible." In 1938, Rudolph Borchardt, a Jewish writer persecuted by the Nazis, wrote this of native plant advocates like Jensen: "If this kind of garden-owning barbarian became the rule, then neither a gillyflower nor a rosemary, neither a peach-tree nor a myrtle sapling nor a tea-rose would ever have crossed the Alps. Gardens connect people, time and latitudes . . . The garden of humanity is a huge democracy. It is not the only democracy which such clumsy advocates threaten to dehumanize."

was imported to New York in great numbers by Frederick Law Olmsted, the planner and landscape architect. Many people consider it ugly, but I respect it for its ability to grow in the most adverse-seeming conditions. In the alley in the spring, the leaves of the tree of heaven spread out over the rat holes in Edens Alley, and although the leaves don't actually *look* green, on account of the fluorescent glow of the one streetlight, you are, even at night, reminded of the color green.

And then at last, at the very back of Edens Alley, in the intimate intersection it makes with Ryders Alley, in the corner, there are the large doors, the back entrance to a gourmet supermarket. Soon I would watch as many loads of garbage were carried out when these doors, these gates to Edens Alley, did at last open wide and the garbage was dragged down the alley's slimy surface. In Edens Alley, the cobblestones seem to be older than those in Ryders Alley, though I can't see why they wouldn't have been put in at the same time. Edens Alley's cobblestones look like bad teeth.

Before I complete my description of this nondescript place where I watched rats roam, I should mention that there were things that I did *not* notice at first. One of those things was that the alley was on an incline, that you had to walk uphill when you walked into the alley. I didn't notice the extent of the incline for a long time, until that winter, actually. In fact, I spent months digging and digging into Edens Alley, finally seeing what I had not initially seen, what I had not noticed at first. Again I turn to Thoreau, in *Walden:* "My instinct tells me that my head is an organ for burrowing, as some creatures use their snout and fore paws, and with it I would mine and burrow my way through these hills. I think that the richest vein is somewhere hereabouts; so by the divining-rod and thin rising vapors I judge; and here I will begin mine."

I did notice the giant hole very early on, however. I describe the hole as giant because I only ever went to the rat alley at night, when the rats were out. At night, I couldn't see the bottom of the hole, even when I shined a flashlight down. And besides, when I first started visiting the alley, I didn't like looking in the hole. I would lean over the railing that fences it off and peer down and get a little freaked out. After all, I had no idea what was down there.

WHY IS THE ALLEY NAMED what it is named? What are the historical origins of the names of these little hidden routes? And who cares? As it happens, the history is vague and incomplete, suffering from the alley's obscurity even among obscure alleys.

When I began my time in the alley, I searched for clues in the books that recount the origin of New York City's streets, but there was nothing. A book called *Nooks and Corners of Old New York* described one particular obscure alley as, in the words of the author, "so completely forgotten that in years its name has not even crept into the police records." But it did not even mention Edens Alley. In contrast, Theatre Alley is a star among unknown places, an often-mentioned secret. For more clues, I went back and looked at all of the oldest maps of the city. I saw the city's streets in their primordial moments, the lanes born of Indian trails, which had likely been adopted from the trails of animals. I saw the lonely-looking lanes turn into little dirt roads and then into cobblestone streets and then into the similarly sized strips where millions of people would walk every day and rats would scramble at night. History, it seems to me, transmits more clearly in the intimacy of the still-original streets of New York's crooked-shaped downtown, or in the snugness of any old downtown: think of the mews in London's City, of the shoulder-to-shoulder lanes of Edinburgh, of narrow streets of the Trastevere in Rome and the little bend in an ancient alley that marks the spot where Caesar himself was killed, a crook that marks the course of history. I saw Edens Alley as a crack in the city's early map, an unnamed little pass.

Ryders Alley first shows up on the old maps around 1740, though unnamed. It could be older. In records of the proceedings of the city from just shortly after it had been taken over by the Dutch, there is an order, on April 30, 1672, for a man named John Rider to pave his street, and maybe some of those stones that the rats I observed running across are the ones he used—though most accounts report that cobblestones were first used in 1860. There was a John Rider who was a prominent English attorney at the time, and it seems possible that the alley is named for him. But there is no Rider Street on any known maps until 1755, when it is shown as an L-shaped street but not identified by any name. It appears as Rider Street in 1797, then over the years is variously referred to as Riders Street, Ridder's Street, Rudder Street, and then, at last, as Ryders Alley.

The history of Edens Alley is similarly indistinct, although after digging through the library, I did learn a few things about it. I learned that a man named Eden lived on Ryders Alley after Rider (assuming Rider ever lived there). Medcef Eden was a prosperous New Yorker. In fact, his was one of the most expensive houses in the city. Ryders Alley appears to have been his home base. On Rider Street, Eden had, according to his will in 1798, "five houses and lots of ground on Rider Street; also four houses and lots in the backyard; also ten in the front yard and also four in said front yard at the time of making this my last will unfinished . . ." In a city directory of 1833, the smaller portion of Ryders Alley had become known as Edens Alley. It reverted to Ryders Alley at some time in the early 1900s, until a few years ago when a historic street marker replaced the nonhistoric street marker, the new street marker using the old name.

In 1920, an account of old New York said this of Eden's life: "[V]ery little is known." What *is* known is that he came to New York from Yorkshire, England, probably just after the Revolutionary War. His wife's name was Martha, and he had two sons, Medcef junior and Joseph, and four daughters, Sally, Anne, Elizabeth, and Rebecca. He was one of the last friends of Aaron Burr, the Revolutionary War colonel, banker, senator, and vice president under Jefferson, who did not have very many friends in New York City. Burr had killed Alexander Hamilton in a duel and then, a few years later, was tried for treason on the suspicion that he was going to set up an independent state in the Southwest. (After his trial, New Yorkers never trusted Burr again; he was thought of as a rat.) While Eden referred to himself as a brewer, his legacy was in real estate. He was one of Manhattan's first real estate moguls, in fact, and he preyed on people more vulnerable than he. As it turns out, this early inhabitant of my rat alley acquired his land in little bits at a time from people who went bankrupt, eating up other people's fortunes.

If the rat alley, Medcef Eden's former main residence, is one of the most obscure pieces of land in Manhattan, then the farm he owned ended up being the opposite. Eden's farm was located outside the old city, approximately four miles from Edens Alley, along the road that was then called Old Bloomingdale but is now known as Broadway. It consisted of about ten acres of stream-crossed, green meadows, an orchard, two small houses, and two barns, and it was bounded on the north by a path called Verdant Lane. The farm was sold a few times by the Eden family, who ended up having their own bankruptcy problems, and eventually it wound up in the hands of John Jacob Astor, who sold it himself. Around 1910, the Eden Farm was developed into Times Square. As the lost sibling of the *least* obscure patch of land in America, the rat alley sometimes seems to me the most forgotten place in the city—a lost tide pool on the shore of a great ocean. On the other hand, sometimes I just think of it as an alley, filled with a lot of rats.

*Chapter 5*

# BRUTE NEIGHBORS

IN THE CITY, rats and men live in conflict, one side scurrying off from the other or perpetually destroying the other's habitat or constantly attempting to destroy the other—an unending and brutish war. Rat stories are war stories, and they are told in conversation and on the news, in dispatches from the front that is all around us, though mostly underneath. If you ask people about rats at watercoolers or cocktail parties or over cake at the birthday parties of small children, then you may hear the story of the bartender who had a rat fall from his thatched-roof-decorated ceiling onto his bar where he quickly clubbed it or of the busboy who used a BB gun to shoot rats out in back of a jazz club on Fifty-second Street or of the graffiti artist who remembered a pack of rats falling down on his head from the top of a dormant subway car or of the woman who was enjoying a cocktail one cool summer afternoon in a cool outdoor café in Greenwich Village when a rat ran across her shoe on its way into a hole in the sidewalk or of the men, women, and children who have watched them intently in the subways, as they run across platforms, sometimes get on board subway cars, then detrain at subsequent stops. A particularly good rat story—and I have heard many—was told to me by a research librarian named Stan, who was, at the time of his rat story, living on the Upper West Side. It goes like this: "I was in my bedroom—I had two roommates at the time, but they were both out for the evening—and I heard a noise in the bathroom. Sort of a rustle-around kind of noise, so I got up and peered into the bathroom, looked all around, and that's when I saw the giant rat running around inside the bathtub. I screamed like a girl and closed the door. Then I poured a glass of scotch to think things over, and I decided my best option would be to poison the rat without trying to confront it. So I looked around for some kind of poison in the apartment and all I could find was furniture polish—it was Lemon Pledge—and also I thought I remembered something about rats liking peanut butter, so I mixed some Lemon Pledge with peanut butter and put it on a tiny piece of cardboard and slid it under the door and then waited. After about fifteen minutes I decided to peer in, and the peanut butter was untouched. So then I looked in the bathroom and didn't see it. And then I looked around and it was up on the sink staring at me. And then I screamed like a girl again and closed the door. I had another scotch and ultimately decided I needed to go in there and confront it, so I opened the door, didn't see it anywhere, looking all around, but there was a towel on the shower-curtain rod blocking a window ledge, and I figured it had to be on the ledge, so I grabbed the towel to pull it down, and I can't figure out to this day if I saw the rat on the ledge or it just dropped down into the bathtub, but now it was running around in the bathtub as scared as me and it couldn't get out—its claws were scraping—and so I decided I could drown it. So I turned on the water, waiting for it to drown, and that's when I had the tragic realization that rats can swim. But then I thought I've got it trapped, so now I just have to kill it, and I went back to the kitchen looking for something more poisonous than Lemon Pledge and I found the Comet kitchen cleanser. So I went back in and the rat was swimming around in one end of the bathtub, and I poured a bunch into the other end of the tub, and it formed this large, scary green pool. The rat swam toward it, and the second it hit the pool, it turned belly-up. Then I realized I had to get it out of the tub, and I didn't want to touch it, so I used a plastic bag and cut holes in it to drain the water, and I ran it out to the incinerator chute."

CITIES ARE LIKE PEOPLE IN that they all have their own rat stories. In Los Angeles, people talk about black rats in palm trees and rats in large and beautiful swimming pools—and I personally know a guy who traps rats at his arid, in-the-hills home and then releases them on the concrete banks of the Los Angeles River, which to someone who doesn't live in Los Angeles looks a lot like a large cement drain. In Boston, many rat stories surrounded the construction of the huge, under-the-harbor highway, known locally as the Big Dig, when it was feared that rats would take over the streets but didn't, thanks to a preventive rat campaign—though to be sure there were a lot of rats. In the seventies, London was gripped with a rat story that had to do with the rat-poison murder of an exterminator. A historic measure of Paris's urbanity has been the description by Parisians of large rats on its subway system, or metro. I recently visited Paris with an eye toward seeing large rats. When I did not, after spending several hours being watched on metro platforms by wary Parisians, I began to understand that being able to readily spot rats in a specific city is an acquired skill, akin to learning a local dialect.

As is the case in all cities, rat stories considered newsworthy in New York usually have to do with lots of rats, those engagements in the saga of man versus rat that are characterized by huge or seemingly huge rat infestations. Few neighborhoods have been without these infestations. A typical infestation occurred in the Flatlands section of Brooklyn over a few summers during the 1960s. "The children can't go to the store and they can't go to the library because of the danger of rats," said a woman who showed up to protest the infestation at City Hall—protests are a feature of city rat infestations. And rat infestations are not merely a modern phenomenon; they appear throughout the history of the city, dating from the Revolutionary War. For example, rats were reported in unusually large numbers all over Brooklyn in 1893; as usual, there was an explanation, a shifting in the city's functions. In 1893, horse-drawn trolleys were replaced with electric trolleys. The absence of horses in the trolley company's stables meant the absence of horse food. The hungry rats marched out into adjacent homes.

Sometimes, several infestations occur in the city at once, or what *appear* to be several infestations—there are, of course, rats throughout New York at any given time—and this results in the perception that rats are gaining ground against humans—i.e., winning the war. When this happens, certain predictable patterns of behavior occur on the human side. First, the city moves to a higher state of rat alert, and as a result New Yorkers begin to see rats where they never noticed them before. Then, whoever is mayor at the time makes numerous statements that seek to reassure the public that the "ugly conditions," to use the phrase of Mayor O'Dwyer in 1949, will be taken care of. "Something should be done," O'Dwyer said in a statement that was followed by the appointment of a citywide rat specialist. In 1997, during a rat alert, the city formed the Interagency Rodent Extermination Task Force. "New York City is about to launch the most comprehensive rat routing in history," one report said. This rat offensive was typical in that the city trapped and poisoned rats until the rat population was reduced but of course not eradicated.

A more recent example of a rat war occurred when rats infiltrated the garbage of the Baruch Housing Projects on the Lower East Side. The rats fed on garbage thrown into a fenced-off area reserved for garbage. There, the rats multiplied quickly and began to make forays into people's apartments. Subsequently, several television news crews filmed the rats roaming the nearby streets. Television footage of rats can be startling, but after you have seen it a few times it all appears the same: garbage backdrop, rat running, rat apparently startled by television camera's lights, rat turning, rat retreating. In the Baruch case, as often happens in cities everywhere, the rat problem fed upon itself, like rats eating rats. There was a rat protest at City Hall. People chanted, "One rat, two rats, three rats, four. Everywhere I look there's more and more." The mayor at the time, Rudolph Giuliani, became defensive about the rats; he complained that his efforts to kill rats had been ignored. "We make unprecedented efforts to kill rats," he said. "We kill more of them than probably anyplace else. We probably lead the country in rat killing." Around the same time, rats were spotted on the porch of Gracie Mansion, the official residence of the mayor. The mayor subsequently increased his attention to rat issues. He appointed a "rat czar," a position that, from a tactical point of view, was not significantly different from the position held by former mayor O'Dwyer's rat specialist. A week later, a city councilmen organized what was billed as a "Rat Summit." At the summit, a rat expert gave a talk with slides ("I'm not going to show any live-looking pictures today because people are often afraid of that," the expert said), and exterminators handed out cards, and a garbage disposal salesman showed off a new garbage disposal, and a man representing apartment dwellers plagued by rats testified about the time he dressed up in protective football gear and went after the rats in his apartment with a baseball bat. The councilman who organized the summit, Bill Perkins, said what will likely be said in one neighborhood or another as long as rat and humans coexist: "This is war!"

What I think of as one of the biggest New York City rat battles ever is the Rikers Island rat battle, which began around 1915 and lasted well into the 1930s. Rikers Island is a little island in the East River at the opening of Bowery Bay, in Queens, just one of many islands in the East River and all around New York. (Even New Yorkers forget the city is an archipelago.) Once, Rikers Island was small and bucolic and green, an eighty-seven-acre patch of land owned since 1664 by a family of early Dutch settlers named Rycken. The city annexed the island in 1884 and used it as a dump for old metal and cinders. It was one of the first designated dumps in New York, a response to increasing problems in the city stemming from the garbage being dumped offshore; shipping was frequently obstructed by floating trash, and oystermen complained of raking dead oysters out of garbage. Rikers Island worked as an antidote to the garbage problem until people began to complain about Rikers Island. Very soon, Rikers Island had grown into a five-hundred-acre island, a mass of garbage on and surrounding the original island, which, in addition to being a dump, was now also home to a prison farm. "Such a mass of putrescent matter was perhaps never before accumulated in one spot in so short a time," *Harper's Weekly* wrote in 1894. One of the complaints about Rikers Island was rats.

Rats from all over the city came to Rikers Island, arriving on the fleet of garbage scows. Within the island was a seventy-five-acre lake of stagnant water, and the rats lived along the shore, feeding on the garbage, drinking in the refuse-infused lake; with its garbage and its putrid isolation, Rikers Island was a rat utopia. An official with the department of corrections at the time estimated that there were a million rats. The rats ate the prison's vegetable garden. The rats ate the pigs on the prison farm. The rats ate a dog that was supposed to kill the rats. The corrections department baited and trapped, but as is often the case in particularly large infestations encouraged by particularly large amounts of potential rat food, the rats bred faster than they could be killed. There was a suggestion that the city bring thousands of snakes to the island so that the snakes could kill the rats, then a suggestion that the rats be killed with biological weapons—the rats would be inoculated with rat-destroying bacteria, via a poison sprayed onto the garbage shores. Neither of those suggestions were acted upon. Then, in 1930, rats from Rikers Island began to swim the river to Roslyn, Long Island, a high-toned summer community. That fall, according to newspapers, the sanitation department used World War I-era poison gas to kill some of the rats. The next year, a Manhattan dentist named Harry Unger organized a hunting party of a dozen rifle-armed men. Unger and his posse were about to invade the island until the city called them off, fearing the hunters might shoot the prison guards or possibly each other. At last, in the spring of 1933, two exterminators—the Billig brothers, Irving and Hugo Billig—had some success when, after supervising the placement of twenty-five thousand baits around the island, they carried off two thousand rat carcasses on the first day. They estimated three million rats were living on Rikers Island. They estimated that they would be able to kill twenty-five thousand rats, and they saw those rats killed as an investment toward rats that would not need to be killed in the future. (It is not known how many rats they actually killed.) "Remember," Irving said, "each female rat can have four litters a year. Every litter contains from five to twenty-one rats. These young rats will have families of their own in four months, and their children will be having other children in four more months. Now, just figure how many rats we have killed by killing twenty-five thousand." The Billigs were successful, in that they reduced the population significantly, though modern studies indicate that Irving Billig probably underestimated the rat's reproductive capacity.

There was a time when the city was so full of rats that the presence of rats wasn't news; the news was the rats' absence. Rather than resembling a guerrilla force, rats were like an occupying army. This was primarily because of the amount of garbage in the city that rats were able to live on in the middle of the nineteenth century. "Besides the natural accumulation of filth on the streets from the dung of horses and other animals, there are vast collections of refuse matter—offal from houses, peeling of potatoes, the refuse of cabbages, and all those things which the ragpickers and hogs do not carry off—they are allowed to accumulate in very large quantities," said Professor Alfred C. Post, in testimony to the New York State Senate in 1859. Rat news at the time tended to be about new types of extermination, or exterminators who did notable work with snakes or ferrets, or unusual rat-related events. Sometimes macabre rat anecdotes were reported; in the 1880s, rats ate bodies kept at the coroner's office on Chambers Street. But frequently, rat anecdotes were almost cheery. In July of 1897, Peter Drapp, a florist on the corner of Thirty-eighth Street and Fifth Avenue, attempted to kill a rat with a pair of scissors, and missed, occasioning a long prose piece entitled HARPOONED A POLICEMAN. "Like Apollo of old, whose bad aim with a discus gave Mr. Drapp the hyacinths he sells, Mr. Drapp threw wild," the *Times* wrote.

The majority of New York's rat news has mostly had to do with death—primarily, the death of humans by rat poison. A lot of rat poison deaths are suicides. In 1886, when Joseph F. Conway, an unemployed trolley car driver, who had emigrated from Ireland to Kips Bay, killed himself with rat poison after other methods failed, a newspaper headline said: FIRST ROPE. THEN RAT POISON. But there were many accidental rat poisonings. For example, Howard Mettler arrived home from work in 1899 and ate a pie that his wife had made. After finishing, he told his wife that the pie tasted strange. She then told him that she had made it for the rats in their cellar. Mettler immediately called a doctor, who gave him emetics, which he took until two in the morning, when he died.

In 1913, in a three-story town house on West 103rd Street and Manhattan Avenue, a woman was accidentally killed by an exterminator named Christopher Walden. Walden went from room to room in the empty house, and in each room he shut the windows and placed a little pail into which he poured cyanide of potassium. Then, Walden poured sulfuric acid into each pail, running out of each room after he did. The rooms filled with deadly gas. He locked the front door. At the end of the day, the exterminator returned to the house and, with his face covered, opened the windows, until he came to the second floor, where he found a tenant, who had let herself back into the building during the day and climbed to her floor and then died on her bed. The woman, Kate Calder, was sixty years old and she had moved to the apartment six weeks before with her husband, William Calder, a tailor. They had moved to New York only after their home in Massachusetts had been destroyed by fire. Walden, the exterminator, was not charged, but I can only imagine how horrible he must have felt.

THE GENERAL SANITARY CONDITION OF America's cities changed as the twentieth century began. In New York, garbage began to be collected, for instance, and the city hired a man to haul dead horses from the streets. Consequently, the tenor of the war changed, as rats were slightly more scarce in some areas and, thus, a little more newsworthy in their appearance. After World War II, newspapers still ran stories about rat infestations in various neighborhoods, but the rat stories that were given the greatest prominence described rat infestations in high-income neighborhoods. They are reported with shock.

Just after New Year's Day, in 1969, for example, rats were spotted on a swanky stretch of Park Avenue—in particular, in one of the traffic medians that in the spring are decorated with tulips, the bulbs of which rats like to eat. "The rats, at times numbering in the hundreds, according to witnesses, have drawn early evening crowds of curious spectators to the island that divides the north and south roadways between 58th and 59th Streets," the *Times* wrote. The report continued, "Some of the bolder rats, they said, even crossed Park Avenue recently to forage in the sidewalk trash baskets near Delmonico's Hotel, 502 Park Avenue, at Fifty-eighth Street." At the time there were also huge rat infestations in Harlem, the Lower East Side, and Brooklyn; in 1969, the city was working to exterminate rats in an area of 1,600 blocks in predominantly low-income neighborhoods. But the Park Avenue rats were followed with great scrutiny. Most of the rats there were exterminated in two weeks. The *Times* called the Park Avenue rats "refugees from Harlem, where no one in authority seemed terribly concerned about their deprivations." Of course, students of rat behavior will tell you that the rats were *not* refugees from Harlem. Harlem is a mile away. The rats were Park Avenue rats. Still, the notion persisted that rats somehow didn't belong on Park Avenue, that Park Avenue was not their habitat. A woman living near the infestation said, "The idea of rats crawling around on children in the ghetto really hits home when you see them on Park Avenue."

The discrepancy between people who had rats and people who did not was underscored in 1959. It was a time when Americans and New Yorkers were thinking pretty highly of themselves, when people on Park Avenue felt safe from rats. It was during the Cold War, and Soviet officials were in Manhattan visiting a technology show that highlighted Soviet inventions. A headline in the *Daily News* boasted U.S. EXPERTS WANDER AT RED SHOW & WONDER AT NOTHING. The same week, however, a three-month-old baby died in Coney Island. His mother had heard him crying in the night and thought he wanted a bottle but soon discovered he was being bitten by rats. The baby had been sleeping in a carriage in the kitchen, and he was bitten repeatedly. Between January of 1959 and June 1960, 1,025 rat bites were reported in New York, twice as many as reported in the next ten largest U.S. cities combined. Sixty thousand buildings were identified as rat harborages by the city's health department—buildings constructed before 1902 that had been designed to house a few families and were now housing dozens. In 1964, nine hundred thousand people were reported living in forty-three thousand old law tenements. And then, rat-wise, things got worse. After the demolition of many of those buildings, in the pits of garbage-injected rubble, rats multiplied. The situation was described in one account as a "rat paradise."

The following year the *Daily News* began a campaign that was intended to rid low-income neighborhoods of rats, as well as sell the *Daily News*. The campaign featured what is perhaps the greatest number of rat stories ever published in one newspaper at one time. The campaign was entitled "The *Daily News*' do-it-yourself campaign to rid New York City of 8 million rats." The *Daily News* paid for teenagers to be trained in rat extermination; *News* employees handed out thousands of pounds of poisonous bait from what the *News* called "anti-rat stations," which were often *News* trucks filled with poison. For several weeks during the summer, the paper ran rat story after rat story, written in that fevered language of war. Some headlines:

THE NEWS OPENS ALL OUT WAR ON RATS

1,000 TEENERS WILL SPEARHEAD ASSAULT ON RATS

RAT-BATTLERS TAUGHT TO KILL HIDDEN ARMIES OF RODENTS

THIS IS IT! WE PASS AMMO TO TROOPS OF THE ANTI-RAT WAR

WAR IS ON! RAT-BATTLERS OPEN ATTACK IN E. HARLEM

CITY'S RATS ARE LIVING IT UP ON MEALS THAT WILL BE FATAL

The *News* published a clip-out rat reporting coupon that looked like this:

**REPORT THAT RAT!**

**I have seen rats at the following locations:**

________________________________________

________________________________________

________________________________________

________________________

**Signed**

***(Clip and mail to* THE NEWS, *220 E. 42d St., New York, 17, N.Y.)***

***Paste this on a postcard for easy mailing.***

Often dispatches from the rat war are just rat observations in the field, undertaken by the local citizenry—little accounts that turn the citizens of New York into nature writers, even poets.

From a letter to the editor of a newspaper on October 6, 1905, signed "Precautious": "During a seven minutes' walk from 116th to 118th Street in Morningside Park yesterday, I saw crossing or attempting to cross the alley five rats, one of them of uncommon size."

From a doorman on the Upper East Side, in 2002: "They are huge, big five-pounders or more. They come out right in front of you and you've got your breakfast in your hand, and you drop the bag and you start running."

From a news account in 1889, based on the observations of the night watchman at the ruins of the old Washington Street Market, an open-air market that had just been demolished that year—the area was said to be alive with rats: "He heard a scuffling noise under the boards at the end of the pier, and, holding his lantern over the string piece, he bent his body cautiously forward and saw an eel trying to lower itself into the water, but badly handicapped in its efforts by a rat that was hanging to its tail. The eel, after the manner of its kind, had crawled from the water in the night to search for food, and had found the rat. The rat, too, was looking for a meal, and was glad to meet with the eel; but when each discovered that its intended supper wanted to eat it, a fierce encounter ensued. Judging from the appearance of the combatants when the watchman saw them, the rat was getting the best of the contest, and, though somewhat lacerated, was fighting with the unbounded confidence of assured success, while the eel was evidently running away. The light of the lantern startled the rat, and, opening his jaws, he allowed the eel to escape into the water, while he went sulkily back into the darkness. The watchman thinks the eel was a yard long, and the rat, he says, was as big as a well-grown cat."

And finally, a firsthand report from a man running down the steps of a tenement in Brownsville, a burned-out and at that time almost uninhabitable section of Brooklyn, which Mayor John Lindsay toured in 1967, the year there were riots in many American cities, though not in New York, in part because Mayor Lindsey was in the streets; the man had a just-killed rat in his hand: "*This* is what's happening, baby!"

One of the most amazing rat skirmishes ever happened downtown in 1979, when a woman was attacked by a large pack of rats.

It happened on a summer night, just after nine o'clock. The woman was described by the witnesses as being in her thirties. She was on Ann Street, right near the entrance to Theatre Alley. Judging from the various accounts, she seems to have been approached by the rats as she was walking toward her car. She also seems to have noticed the rats coming near her, their paws skittering on the street. Witnesses said the rats swarmed around the woman. One climbed her leg and appeared to bite her. The woman screamed. A man nearby ran to help the woman, taking off his jacket and waving it in an attempt to scare the rats away. The man told police that the rats appeared undeterred by anything he did, and in a second they began to climb up his coat. Seeing this, the man ran to a phone and called the police. By now, the woman was "in a state of hysteria," according to another witness. She staggered to her car, which was parked a few yards away. The rats followed her. She got in, closed the door. Now the rats were climbing on her car. She was screaming when she drove away. The woman was gone by the time the police arrived, but the rats were still there, scurrying through the street and into Theatre Alley and into their nests in a lot on Ann Street, just around the corner.

The lot where the rats lived had been a bar that had blown up nine years before. Ryan's Café was an old stand-up bar in a building that was thought to date back a hundred years. It exploded at two in the afternoon on December 11, 1970, shortly after some customers had complained of smelling smoke and two workers went to the basement to turn off a water heater. Eleven people were killed and sixty were injured. After it exploded, people were clawing through rubble, bloodied and crying. In City Hall, across the street, the mayor thought he'd heard thunder. Police helped a man crawling from the rubble; his clothes were smoldering. A nurse ran over from the construction site, one block west, where the World Trade Center was being built. "Dirty white smoke curled high across Broadway, wreathing the spire of St. Paul's to the southwest, drifting across the ornate front of the Woolworth building to the northwest, finally dissipating around the skeleton of the World Trade Center far to the west," a newspaper reported, adding, "The explosion drew hordes of spectators."

A short time later, the old building was demolished. Rats began to nest in the resulting hole in the ground, feeding off the garbage in Theatre Alley. They thrived in the forgotten-ness. Then, in 1979, there was a tugboat strike. Whereas the city customarily transported sixteen barge-loads of garbage every day to the Fresh Kills landfill on Staten Island, during the tugboat strike the garbage was all backing up in New York's streets. Eventually, President Jimmy Carter declared a health emergency in New York City and ordered the Coast Guard to tug trash barges, but in the meantime trash was tossed into the old lot that had once been Ryan's Café. The old lot that was once ideal rat habitat now was the *ideal* ideal rat habitat. And the strike had gone on for weeks; the streets had not been cleaned for three months. An artist, Christy Rupp, who lived across the street from Theatre Alley, had been hanging up drawings that she had made of large rats, to draw attention to the rat situation. Rupp said she had seen fifty rats in three hours on the night the woman was attacked. "Rats are everywhere," Rupp said shortly after the woman was attacked by rats. "They are very successful urban animals. They are king in New York." (After the Ann Street incident, Rupp went on to create plastic sculptures of rats—rats standing, crouching, and rolling over.) Meanwhile, the identity of the woman who was attacked was never discovered. The police called area hospitals but no one reported being attacked by a pack of rats.

Randy Dupree was the head of the city's pest control bureau in 1979, and on the night of the rat attack near Theatre Alley, he was in upstate New York, at a rat control convention. After the rat attack, the mayor, Ed Koch, telephoned him and ordered him to return to the city immediately. The next day Dupree was watching the rats run back and forth to Theatre Alley for food. Initially, Dupree estimated one hundred rats were living in the lot. The bureau began trapping and poisoning. They used peanut butter laced with poison. Workers in the pest control bureau crawled down into the hole and beat the rats with shovels and clubs. Dead rats were flying around the hole that was Ryan's Café. Reporters who went down into the hole ran out frightened. The workers dug through foot after foot of pipes and lumber and rotting garbage—two tons of garbage in the end. The first day they killed a hundred rats. Dupree was forced to upgrade his initial estimate of the rat population after they killed a hundred more.

I ran into Randy Dupree at the Rat Summit in New York City a few years ago and arranged to meet him for lunch one day in a restaurant in Manhattan. Dupree had recently retired as head of pest control for the city after twenty years; when we met, he was consulting for pest control companies in the city. "I surveyed every block in New York City for rats," he told me. He still finds himself looking for rats wherever he is, checking the streets, looking around restaurants. "Instinctively, I do," he said. "Runways are always there." He shook his head. "You know, I tried to divorce myself from rats. I said, 'I hate rats. I don't like 'em.' But then, there came a time I said to myself, 'Randy, hey! That's what you do! That's who you are!' Now, any of my friends, they see a rat, they say, 'Where's Randy at?' "

Dupree had no problem recalling the rat attack off Theatre Alley: "I remember that, I sure I do." His explanation of the attack made it seem less warlike and more accidental, a crossing of human and rat paths. "What I think happened is—well, this lady, she got off work at eight or nine o'clock, and it was getting dark and that's when the rats would have started migrating from the alleys. And then, what happens first is, the lady sees the rats. And then the rats see her, and then she starts running and then the rats start running, because look—they're just as scared as she is. And they're there, all of them just running along. And so it looks like they're attacking her. But then because she thinks they're attacking her, the rats get scared and then they actually do."

Dupree has the confident smile of a man who has heard his share of rat stories. "Number one," he said. "Most people exaggerate. You know, the rats-big-as-cats stories. You start to discount them." On the other hand, he has his own rat stories that stretch the limits of the imagination, such as the story about the rats that were discovered in the 1980s in the Mott Haven section of the Bronx, in a place called St. Mary's Park. All of the rats in St. Mary's Park were tested to be resistant to poison—they were so-called Super Rats. "When word got out, you couldn't find a person who would go into St. Mary's Park in the Bronx," Dupree recalled.

As he was telling me about the park, Dupree suddenly noticed a small fly on the table. He stopped talking, watched the fly for a second, then caught it. He looked at it closely in his fingers and placed it in an envelope. He put the envelope in his pocket. "I'll check that when I get back to the office," he said.

He resumed talking. "Anyway, I remember those rats downtown. Yes, I do."

## *Chapter 6*

# SUMMER

THIS WAS MY curious labor: to observe the rats in my alley, to think about these specific rats and savor their natural details, which I did—although I admit I also thought a lot about rats coming at me en masse each time I passed by the scene of the 1979 rat attack on the way to Edens Alley. And, yes, at times the thought of rat attacks made me reconsider what I was doing exactly, especially early on, when my wife, for instance, had not yet come to terms with the idea of me spending my summer evenings with rats. Nevertheless, I managed to set off to describe the movements and details of the rats, to see how rats live. And here at the beginning of my experiment, my thoughts alternated between being trampled by herds of rats and not being able to see any rats at all, much less tell them apart. Would I really see rats? Would they see me?

I went to the alley late each evening, arriving after the office workers had left for home, as people were settled in for a late-night drink at the bars. I would take my position, usually standing at the entrance of the alley. I would then wait—simply and with deliberateness. At first, I relied heavily on my night-vision gear to see the telltale shine of the rats' small eyes, but, as I became more accustomed to spotting the rats, I was able to use standard binoculars. The streetlight allowed me to see the rats with the naked eye, the sickly yellow glare shining in the remnants of a broken windshield, in the plastic newness of each evening's freshly tossed garbage bags. Sometimes, on the very corner of a small triangular plaza in the middle of Fulton Street, I sat on a small camping stool and looked in with binoculars. In the anonymity of a big city's street traffic, I was, as best as I could tell, unobserved by the vast majority of passersby as I watched the rats consume. "They've got four thousand in net capital," a man once said as he brushed against my shoulder, his suit touching my windbreaker, his other shoulder bumping up against the man who was walking along with him, nodding.

I kept notes all that summer, and after a while, I found the alley, aside from its stench, to be a pleasant place, a scene for a kind of rat-related meditation, a place that never ceased to be novel. I looked forward to my visits there, in fact. I merely needed to take a few moments at the start of each evening to acclimate myself to the smell. When I first began observing, I was happy, after just a few minutes of intense, eye-aching scrutiny, to merely see a rat, any rat. But after a few days, I was able to spot rats quickly. Like the birder who returns over and over to the same woods, I grew comfortable in Edens Alley, accustomed.

Once the rats were out in the alley, they moved quickly—sniffing, licking, nibbling, walking easily around empty, beat-up rat poison dispensers, then galloping off in impressive bursts along the cobblestones. I noticed early on that a rat would stick its head into a garbage bag for a number of seconds. I counted off seconds as a rat drank water from a thimble-size puddle in the nooks of the cobblestones: six. I wondered what proportion of their required two ounces of water a day these six seconds represented—like any time-rich nature-watching endeavor, observations beget more and more questions. On another occasion early on, I took a video camera to the alley and filmed a rat running. I expected to see a kind of skittering, a spidery or crablike crawl. But when I analyzed the tape, I was amazed to see that the rat was almost galloping: the hind legs pushing the front legs up and forward, resulting in an elegant midair arch of the rat's body. Given the darkness of the alley, it is difficult for me to say for certain, but it seems likely from the rat-running tape that all of the rat's legs are in the air at some point in the typical rat gallop. I would bet on it.

You might not have known it was summer by looking at the alley. Sometimes the alley seems free of all seasonal indications, or resigned to the bare minimum, at least—the leaves on the ailanthus tree are either on or off. But there are differences in the seasons, just the same. In the summer, the cobblestones are even more slicked with filth, as if they sweat garbage grease. The ailanthus tree's feathery leaves sway just slightly in the warm, nearly still breeze. One might have known it was summer by watching me watch the rats in the alley; I was perspiration-soaked, frequently sucking from a water bottle I'd previously used on a camping trip. Or one might have guessed from the rancid smell that the garbage emitted: sometimes I thought I saw little ripples in the air above the garbage bags, like heat on a highway's horizon, but I would then blink hard and see nothing. In the summer, the alley's bouquet is at its richest, most intense. In the summer, it is like a tidal flat at low tide during a full moon—oh, how it stinks! It was during the summer months that my wife seemed most concerned about me possibly bringing back any diseases from the rat alley. It was then that I was mostly likely to strip down before reentering our apartment, where I then showered profusely because—even though I didn't mention it to my wife—at times in the summer I felt as if I were nature-walking through a petri dish.

During a typical summer evening, I would venture into the alley a few times a night to search for burrows, for tracks and scat. One night, to estimate a rat's speed, I ran alongside an adult as it in turn ran down the sidewalks turning the corner from Ryders Alley into Edens Alley. I put the speed of the rat at roughly six miles an hour, perhaps slightly faster. Mostly, though, I stood on the edge of the alley and concentrated on looking intently.

One of the first things I noticed was that the rats in the alley tended to run close to the walls and curbstones; this observation is consistent with everything the pest control manuals say about rats, their thigmophilic behavior. As for the specifics of my alley, I noticed that rat activities seemed to be centered around two main garbage areas: the Chinese restaurant garbage area on the north side of the alley and the Irish bar's garbage on the south side. The first observation that I made, that I felt was a legitimate *Rattus norvegicus* observation, was this: the rats seemed to stay on the side of the alley that they were eating the garbage on.

As for the total population, it was difficult to get a count of the rats initially. I could see as many as seven or eight at one time, and that number seemed to increase as my observational abilities improved. Judging from the amount of rat scat and from what I had read in Dave Davis's papers, I guessed that there were at least a couple of dozen, which seemed like lot for a small alley.

And yet despite what I felt were noticeable improvements in my rat awareness, I still found it difficult to recognize individuals. When I saw rats, I just saw rats. My instincts told me I had a lot to learn.

A SIDE EFFECT OF MY close attention to rats in my alley was that I was suddenly much better at seeing rats in alleys everywhere. One night, however, when I was looking down into Theatre Alley, seeing the telltale scatter, I met someone who was skilled in rat watching in a way that I will never be—a kind of accidental master. His name was Derrick, and he was hanging out in Theatre Alley. When I met him, I was reminded of a description of a certain species of man that Thoreau calls "wild"—men who, as he put it, "instinctively follow other fashions and trust other authorities than their townsman, and by their goings and comings stitch towns together in parts where else they would be ripped." Thoreau continues: "His life itself passes deeper in nature than the studies of the naturalist penetrate; himself a subject for the naturalist."

I learned a lot from Derrick, though at first I didn't notice Derrick. I saw the rats in Theatre Alley scurrying; I noticed them moving along the base of the old brick wall, saw them hugging the corners as they came out of an abandoned lot. I have to admit that I was wondering if I should have made Theatre Alley the centerpiece of my rat studies—a certain rat envy came over me. But in the end, if there was one thing that meeting Derrick did, it made me thankful for the situation I had.

It was hot that evening, extremely hot and humid. German cockroaches were on the ground, near the trash, but American cockroaches were on the walls of Theatre Alley.* The American cockroaches were big and they were flying—an exterminator said that a flying American cockroach is a rare sight, that it has to be especially humid to see a roach fly in New York. ("They have become what we call domesticated, in a sense, because see cockroaches *have* wings—they used to fly everywhere to get food—but now they don't use them so it's like someone who has been in a wheelchair for a long time and they've come to the point where their legs don't work anymore," the exterminator said.) In fact, I was bobbing and weaving, ducking the flying cockroaches, as I stood there looking down the alley for rats. I noticed a number of stacked-up cardboard boxes, ready for recycling, which the rats were using for cover. I counted six rats on the eastern edge of the alley, and as I was counting, I noticed Derrick, who, in turn, noticed me. He seemed confused by my presence in the alley, perturbed even; he said something to the two people he was standing with. He stopped what he was saying to the man and the woman he was with and, leery, moved toward me. I greeted him. He was reserved. It wasn't until I mentioned that I was looking for rats that he finally brightened. "We got lots of rats," he said, gesturing all around the alley.

Derrick is not tall—about five foot seven—and he is wiry and scraggly. He talks quickly and, as he does, moves his hands. He seemed to be living on the street. He was reluctant to talk about his situation, and I didn't press him. It did come up that he was forty-four years old. He kept a poster of Jimi Hendrix hanging on a chain-link fence that was surrounding a lot filled with the remains of a demolished building: a hole of rubble and garbage on Theatre Alley. Derrick told me that a few months before a man had been filming some of the rats in the

---

* Just as Norway rats did not come from Norway, German cockroaches did not come from Germany; more likely, they came from Africa. In Germany, German cockroaches are known as French cockroaches or Russian cockroaches. In Russia, they are known as Prussian roaches. One theory has it that American cockroaches were brought over from Africa by slave traders. American cockroaches are also known as Croton bugs because they arrived in New York City through the water pipes laid in to carry water from the Croton reservoir system, in upstate New York.

alley. Derrick said that he had been trying to lower the rat population in the alley himself but hadn't had a lot of luck. "We've been trying to get rid of them some nights," he said. He shrugged and shook his head in wonder. "You will kill yourself but you will not kill these things."

He began to describe the various groups of rats that lived in the alley. "I've got a couple of different groups of them." He turned around and gestured with his beer bottle to one side of the alley, behind some Dumpsters, and then up high, on top of a garbage area, where rats were nesting on a corrugated-metal roof—you could hear their nails scratching the tin. He pointed out the camps of rats, which I understood to be individual nests. "I got a whole posse of rats over here," he said, "and I've got some here and then ones over on this side. And we used to have rats over here but they left. I guess they got outnumbered. And these here, they will fight the ones on the other side. They just tell 'em—you don't belong in this area. It's just territorial rights, like a lion would do, you know? Also, sometimes they hiss at each other."

At that moment, a large rat entered the very center of the alley—so at ease were the rats, I later surmised, that some of them did not feel the need to touch walls.

Derrick put down his beer bottle and approached the rat. The rat made a hissing sound. Derrick made a hissing sound in response. He turned back toward me to make a point: "They get on their hind legs," he said, "and they jump like this." He jumped.

The rat hissed. "See, he's trying to tell me that he's not scared. So I, Derrick, say, 'Okay, I got to get a brick for your ass.' 'Cause I gotta teach 'em who's boss in this house."

Derrick hissed again, this time lunging forward and stomping his foot as he did. The rat retreated.

Derrick picked up his beer bottle again, relaxed. He noted that professional exterminators had come to the alley. "The rats look at the traps like they're some kind of a joke," he says. And he recalled a time that there was no garbage in the alley for a couple of days and the rats ran around a little more on edge than usual, exhibiting signs of cannibalism.

"It's just like humans when they don't have any food," he said. "It's like that plane crash where the people started eating each other."

After a while, Derrick went back to the area of the alley where he had just faced off with the large rat. He wanted to demonstrate the prowess of the rat he considered to be the most aggressive. He sought it out and then challenged it, with stomps and sounds.

"*Shhhhhhh-oooo! Shhhhh-oooo!*" Derrick said.

The rats squealed in unison.

"Do they normally respond to that?" I asked.

"I've got one or two that will stick around. The rest will run. But I got one or two really big ones. I guess they're the elders."

One of the rats was climbing up and down a fence toward the corrugated-tin roof. "They're getting innovative," Derrick said. "They're getting too damn smart for their own good.

"Look at that!" he went on. "Look at his imagination! So, okay—so he's learning too much for his own good. It's still survival of the fittest. They're cunning little animals. That's why I don't trust them."

I felt hopeful watching Derrick with the rats—hopeful that I might eventually understand rat behaviors in my own alley, I mean. But at the same time, I was worried about Derrick's living situation. I'd read about a lot of people falling asleep and being eaten by rats in alleys, for instance: the rats are drawn to the smell of food. I was just standing there watching Derrick, feeling a little dumbfounded, when one of the guys he was drinking beer with called out to him:

"Hey, rat man!"

Derrick turned around and looked at me. His eyes were beaming as he shook his head. "It's a very interesting world, isn't it, the lower echelon?"

I ONLY STOPPED BY THEATRE ALLEY to see Derrick a few times that summer. I had my own rats to attend to, my own alley to study, and I didn't want to bother him too much; he always seemed to be negotiating with people in the alley, his base. After a while, I had the feeling that I was beginning to annoy him; I was feeling like a pest.

Even with his expertise in rats, he was understandably only so interested in them. Also, whereas I was entertained by the rats' movements, like a hermit in the woods watching squirrels, he was *living* with rats, which was a whole different story.

Nevertheless, when I saw him one night late that summer, I waved down the alley, and he squinted to see me in the alley light, then came out and greeted me with a hug. I was with two friends, Matt, a poet, and Dave, a painter. I suppose the mere fact that I was in the company of two friends itself proves that I wasn't actually like some kind of hermit when it came to my rat studies. Sometimes my friends came along to help, to observe, to share in the ratness of it all. (Believe it or not, I sometimes had to turn people down when they requested to come to the alley.)

Derrick was also with a companion—a tall guy, quiet, laughing occasionally—and he was holding a beer bottle and weaving a little. I noticed that Derrick had cleaned up his area of the alley using the brooms that were laid next to his chair and his portable radio. I said the alley looked clean and pointed out that it seemed as if there were fewer rats around as a result. Derrick disagreed. He took me back in the alley and shouted and stamped and made the squealing ratlike sound that he makes. Rats came racing out.

This time there weren't just five or six rats scurrying around. This time there were lots of rats—dozens of rats. I would say there were close to one hundred rats. (Don't think I'm just saying "close to one hundred rats." I arrived at this number carefully, comparing notes with Matt and Dave that evening and factoring in possible hysteria. I'm as certain of it as I can be.) All of the rats seemed to have joined forces and come out en masse, as one pack of rats. It seemed reminiscent—to me, at least—of the rat situation involving the woman in 1979 who either was or merely thought she was being attacked by rats right in this location, on the corner of Ann Street and Theatre Alley. I'd never seen anything like it, except in movies, and in movies, I happen to know, they mostly use trained nonwild Norway rats.

"See," Derrick said, as rats came rushing out everywhere. "I got 'em trained. They're just like any animal. If you do something long enough, you've got 'em trained."

The rats were running from all the nests he had described, and to make a point that bears repeating, there were a *lot* of rats. In addition to the running, the rats were screeching, screaming, and making other noises. Derrick was saying, "See? *See?*" I was nodding, and looking down at my feet. I was trying to simultaneously count the rats and stand still and stay out of the rats' way.

The rats moved in the shape of a mob, a herding mass, with rat trying feverishly to pass rat, some not passing, some falling back, some climbing past the others. Matt and Dave and I gathered close together, as if we were about to be burned at a stake, and we watched in panic-stricken amazement, deciding instinctively, I think, that it was better to stand very still than to run.

Fortunately, instead of heading straight for us, the rats suddenly veered off toward the wall of the alley, at a point a few feet before the alley wall ended and the vacant lot began. The lot, as I have already mentioned, contained the remains of a demolished building; it was a vortex of rubble. The rats were racing, at high speed, and yet as they turned and came to the wall, they formed a single-file line, seemingly meticulous, purely straight: I thought of commuters massing on the street and then filing single file down the narrow steps of a subway entrance, of spectators filing out of a baseball stadium. The rats turned the corner of the alley wall and headed down into the lot as if they were flowing out of a funnel or a spout. "A rat faucet," Dave, who was standing very still next to me, remarked. From there, the rats dispersed, skittering into the rubble hole, turning from rat to moving patch of darkness to shadowy blur as they scrambled down and down and, one by one, disappeared into the hole.

After the rats were gone, my skin was tingling. I don't know about their skin, but Dave and Matt seemed pretty shook up too. Dave said, "Jeez." And Matt was saying he couldn't believe how many rats he had seen. I had a couple of out-of-body experiences, in which I communicated with my body asking it what in the hell it was doing there. We looked over at Derrick, who was watching the rats as if he were a shepherd. One rat had stayed back, undeterred by Derrick's calls and commotion.

"See that! See that *one!* I always get one. It must be an elder. I always get one who will stand there, and that's what upsets me. 'Cause if it doesn't move, I'll have to kill him, and they're my babies. I call 'em my babies, anyway."

He was serious about this. He turned toward us. "Some of these rats are too smart for their own good," he said. He looked me right in the eye. "You know what I mean?"

I shook my head. "Yes."

I was interested in leaving at that point, so we all said good-bye to Derrick, shaking his hand. I walked the few blocks over to my rat alley to take notes—excited, in fact, by everything that Derrick had showed me. But then, about half an hour later, we saw Derrick again. We saw him all of a sudden. He had popped up out of a little crawl space between an old power plant and a Thai restaurant, a little spot on Gold Street, where he apparently slept. Derrick seemed more shocked than we did, and we were pretty shocked.

"Hey, are you following me?" he asked. He was angry.

"No," I said. "We're just looking for rats in this alley."

"You're following me!"

I felt like a jerk. "No, really," I said.

He finally calmed down after that and began talking about other places he had seen rats downtown. He said there were rats under the Brooklyn Bridge, in a spot where he had once often slept: "Me and my wife, we were living there. She died there."

Derrick looked at me again, right in the eye. "You know, some guys are too smart for their own good," he said. It made me want to go home and never look for another rat. He was still staring at me, and he said it again. "You know, some guys are too smart for their own good."

*Chapter 7*

# UNREPRESENTED MAN

WE ARE FOREVER complimenting the so-called Great Men, forever scrutinizing their glory-gaining actions, the endeavors that light up the past like torches in a great hall. In "Representative Men," Emerson writes: "Nature seems to exist for the excellent. The world is upheld by the veracity of good men: they make the earth wholesome. They who lived with them found life glad and nutritious. Life is sweet and tolerable only in our belief in such society; and, actually or ideally, we manage to live with superiors. We call our children and our lands by their names. Their names are wrought into the verbs of language, their works and effigies are in our houses, and every circumstance of the day recalls an anecdote of them." It is likewise that the nature in which rats exist calls to mind those lives that are not recalled and honored, whose careers are not reexamined in histories—the lives that seem to be unnatural and even ratty or at least low-down but are not, actually. For isn't it the case that the life that exists in the sulfurous swamp, in the stink-laden dreck at the bottom of a bog, is as fertile as that in the alpine waterfall, if not more so, even if it is less classically picturesque? When we look at rats, we are thus compelled to look at the history of the lives in their midst, to search for the Unrepresented Men. To quote Emerson yet again: "Each man is by secret liking connected with some district of nature, whose agent and interpreter he is; as Linnaeus, of plants; Huber, of bees; Fries, of lichens; Van Mons, of pears; Dalton, of atomic forms; Euclid, of lines; Newton, of fluxions." So it is that Jesse Gray is, in my rat-interested mind, the agent and interpreter of rats. Jesse Gray was a tenant organizer in Harlem in the early sixties who screamed and scratched and hissed at the people in charge of the city but didn't really have much luck until he used rats.

Jesse Gray was not tall but he looked a lot bigger than he was, especially in the winter, when he was always bundled up in layers and running from a picket line outside of City Hall or outside the police department or else stopping by the home of someone who had no heat. If you knocked on his door and the assistant answered and you said, "Mr. Gray?" the assistant would reply, "No, I'm Mr. Brown." People described him as nervous or agitated, as "a formidable bundle of energy," in one reporter's words. His distinguishing characteristics included a small mustache and two large bucketeeth, which jutted out when he flashed the wry smile that was indicative of a sarcastic sense of humor. Once, after he was found guilty of obstructing the police as they attempted to evict a family from a rat-infested apartment, a newspaper reporter asked if the jail sentence would curtail his operations on behalf of renters. His face lit up and he smiled his big, toothy grin. "Rather, it will serve to increase them," he said.

Gray was born in Tunica, Louisiana, a little town sixty miles up the Mississippi from Baton Rouge, one of ten children. He attended college on and off for a few years, then worked variously as a merchant marine, a short-order cook, a waiter, and a tailor. In New York, he worked on the docks for a while. "My real school was the waterfront and the union," he used to say. In the winter of 1952, when his heat went out in the Harlem apartment where he was living with his wife and children, he joined a tenant organization, called the Harlem Tenants Council. Asked about his inspiration for organizing people on behalf of their rights as citizens, he said, "I was cold."

For years, Jesse Gray worked out of an office on 117th Street, a neat and sparsely decorated little hole-in-the-wall that was just off Fifth Avenue in Harlem. The history of Harlem is, like that of many New York neighborhoods, the history of migrations and change, and one way to sum it up would be like this: Harlem was first settled by Dutch farmers, who named it Nieuw Haarlem and prized it for its remoteness. When the farmland failed in the 1880s, the Dutch were replaced by Irish squatters, who, when brownstones were built, were replaced by European Jews looking to escape the crowds of the Lower East Side, who, in the 1920s, moved to the Upper West Side to escape the overcrowding in Harlem, where they were replaced by African-Americans, who found less racism in the neighborhood than in other parts of the city and also cheap housing. Harlem grew to be a cultural capital, until, first, the Depression destroyed its economy, and then, problems such as heroin addiction ravaged its inhabitants, so that eventually 117th Street, like the bulk of the streets in Harlem, was in decay. Jesse Gray's office was once described as "a tiny oasis of order and cleanliness in one of the most appallingly filthy blocks of Harlem."

Staffed by scores of volunteers, Gray's operation was continually on the verge of bankruptcy. He worked for ten years, agitating relentlessly against landlords without much success, the big problem being lack of sufficient confidence amongst the renters to undertake rent strikes—people were afraid to ignore the eviction notices. In 1959, Gray organized a small rent strike that failed. Then, in 1963, Gray organized the first rent strike utilizing rats. This time renters in 250 buildings went on strike, in an area bounded by 118th Street and 125th Street and Park and Eighth Avenues—thirteen thousand people who were "outraged by their own misery," as *Ebony* magazine wrote. Photographs of the buildings on strike showed ramshackle apartments with hole-riddled windows that often did not close, with caved-in ceilings, with walls shedding plaster. A portrait of a family in *Ebony* showed a group of children huddled under blankets during the day, seeking shelter from the winter wind in their bedroom; tenants described heatless, cockroach-filled conditions. During this second strike, Gray allied with local church officials, with labor leaders, with Harlem politicians—he organized the ghetto. "Bring a rat to court!" Gray told the tenants. They had ample opportunity to do so. According to Health Department statistics, one half of the housing in Harlem was rat-infested.

People brought dead rats and live rats. People dangled rats by their tails for the newspaper photographers; people displayed rats spread out on newspapers, like fresh fish they'd bought at market. People wore rubber rats pinned to their jackets. People testified about rats. "The rats are part of our family," one woman said. "I've seen kids try to pet them," said another. Tenants brought rats to civil court and they brought them to City Hall.*

Gray's second attempt was helped by that fact that the black community in Harlem was emboldened by the recent gains of the civil rights movement in the South, as was Jesse Gray. And this time when he asked the rent strikers to ignore the eviction notices, they did. "Our most important aim is to give the people a consciousness of their rights," he said. "Then landlords will wake up *down*town and come *up*town and see what is happening." Gray equated the tenants to *Rattus norvegicus*—and it is one of a very small number of pro-rat comments I know of in the annals of New York City history. "The tenants are like rats now," he said. "Rats feel their power, and they come out in broad daylight and just sit there. Once the tenants feel *their* power, they stop running, they're not afraid anymore. We've shown them—and they see now—that they have rights whether they live on Park Avenue or Lenox Avenue."

As the winter went on, the strike doubled in size. Gray and the volunteers of the Community Council urged the city to take over the dilapidated buildings. Gray called for "a mass rehabilitation of the ghettos." The courts sided with the rent strike; a judge ordered repairs. To this day, it's the largest rent strike the city has ever seen. In January of 1964, the strike spread from Harlem to the Bronx and the Lower

---

* My father used to own a printing shop down near the South Street Seaport on a part of Pearl Street that is no longer there, and months after I had begun researching Jesse Gray and his rent strikes, I learned that my mother had been walking to my father's print shop one morning when she passed a rent strike demonstration at City Hall. She had no idea what the demonstration was about until she got caught up in it. The police began rounding people into paddy wagons. The next thing she knew she was being arrested, though the police released her when she explained that she was just on her way to work. She also told them she was pregnant. In fact, she was pregnant with me. Needless to say, I was pretty excited to hear that I was once sort of arrested during a Jesse Gray rent strike demonstration.

East Side, including Hispanic neighborhoods. Now, at protest rallies, the signs that said NO RENT FOR RATS and FREEDOM NOW and JAIL THE SLUMLORDS were accompanied by signs that said LAS RATAS. Newspapers reported on rat bites that would not have been deemed newsworthy before the strike. "The rat seemed afraid of no one," said the mother of a five-year-old boy who was bitten on the face, a day after their building had joined the rent strike. (The landlord in that case responded, "What can I say? These things happen.") At a rat-oriented rent strike rally, Gray shouted, "Rats are eating up this community. We want emancipation now from rotten landlords!"

Gray and the Community Council held planning sessions in the crowded community centers and sometimes in jails: a photograph taken at the time shows Gray in a tattered suit and tie in a jail cell with his similarly attired colleagues surrounded by benches full of drowsy men. Then, Gray and his colleagues raced out to apartments, to keep the rent strikers from being evicted—with shouts, with court documents, with barricades, with hurried heaps of worn-out furniture. One day in January 1964, Gray was at the apartment of a man named Luther Brown on West 118th Street. Gray was with some students from City College and a few members of the Northern Student Alliance—that people were forced to live amongst rats was seen as an injustice worth fighting for by student activists at the time. In the hallway of the apartment, there was a scuffle with some city marshals when Gray and his entourage would not leave the apartment. Everyone was arrested—Gray was always scrapping with the police and with City Marshal Henry Lazarus. "The tenant had no knowledge of the impending eviction," Gray said, as he was led in handcuffs down the narrow, rickety steps. "The police department reacts with great speed to uphold the law for the slumlords. We ask them to show the same speed in arresting the slumlords and protecting the people."

After being booked at the West 123rd Street precinct, Gray was taken to Harlem Hospital where he was examined for injuries resulting from a police officer's foot in the back. The family was evicted. Homer Bigart, the *Times* correspondent who became famous covering the Vietnam War a few years later, covered Marshal Lazarus as he broke open the door of apartment 4E with a crowbar—and it seems appropriate that a future war correspondent was on hand in a rat-pit-like Harlem apartment. "Mr. Brown and some friends retreated to an inner room, throwing up successive barricades of furniture to stall the eviction," Bigart wrote. "After the City Marshal and his men had removed the furniture from the apartment, Laura Brown and her five children, who share the place with her brother, Luther, re-entered to salvage any remnants of the household. They found only a bassinet and a broken mirror."

We know where George Washington dined when he was in New York, in addition to where he slept, as he just barely stayed alive against the British, or we think we know—even the lore of Representative Men is murky. And then there are the lives—lives that are, though not George Washington's, perhaps in some crucial way just as historic—whose record, if it exists at all, evaporates and fades, even now, as garbage putrefies and is dispersed through the city by the tread of shoe, the ruts of a radial tire. So Jesse Gray and the rent strikers have faded—even though the rent strike worked.

The city passed a $1 million rat extermination bill because of the rent strike, for example, and financed the repairs on dozens of buildings; in the summer of 1964 only 60 or 70 of the 325 apartment buildings originally on strike were still striking. At a press conference with a deputy buildings commissioner, Gray said he was "well satisfied," though there were still buildings he hoped the city would take over from the landlords. "It's not enough," he said. In 1967, Jesse Gray brought a rat in a cage to Congress, where Southern Democrats ridiculed a federal rat control bill, calling for a less expensive release of cats in urban areas, predicting a "federal ratocracy." As Gray's group was arrested, they chanted, "Rats cause riots. We don't need a riot bill. We need a rat bill." President Johnson secured $40 million in extermination funds. Later on, in 1971, President Nixon proposed cutting federal rat control funding, but reinstated the money, in 1972, after being criticized for the way he dealt with rats.

An even less visible but more significant result of Gray's rat strike was the way in which Gray's grassroots group energized grassroots groups like it all over the city, and possibly the U.S.—one historian wrote that Gray's strike helped spawn the National Tenants' Organization, in 1969. It was the time in America when urban renewal was paving over old neighborhoods in New York in the name of progress and relocating them for the sake of highways, for sterile planned cities that were like laboratory cities, not at all wild. The chief formulator of urban renewal in New York and, because of his influence, in cities all over America was Robert Moses, the city's master builder. Most historians argue that Robert Moses and his destructive policies were finally halted by a liberal elite—groups of upper-middle-class homeowners who organized in Greenwich Village, for instance—but some people say it was the power of the tenants movement that stopped Robert Moses. "It is not too much to say that these sometimes lonely activists . . . shaped the awareness of the dignity and integrity of neighborhoods that would become the most significant ingredient of the community-power movement of the 1960s," wrote Joel Schwartz in a history of the tenants movements. "The tenants forlorn protests . . . helped mold the sense of injustice that would eventually change the course of urban redevelopment in New York and across the nation."

People who liked Jesse Gray liked him for stirring things up. People who did not like him thought he was just stirring things up. During the rent strike, Gray was continually harassed by the police, and he harassed the police back. He led strikes on police headquarters, at one point calling for the resignation of the police commissioner, Michael J. Murphy, who Gray deemed "a servant of the slumlords." Murphy, in turn, accused Gray of creating "an atmosphere of violence." This charge went back and forth, and at one point, Malcolm X, who was attending Gray's rent strike rallies now, argued that he thought the police wanted Harlem residents to resort to violence. "Then they'll be free to put clubs to the side of your head," Malcolm X said. That summer, when a boy was shot by a police officer, there was rioting in Harlem. Gray spoke at a rally at a Mount Morris church on 122nd Street. The theme was "Is Harlem Mississippi?" and when he spoke, Gray, wearing a bandage on his head, said that police had beaten him. "There is only one thing that can correct the situation, and that's guerrilla warfare," he said from the pulpit. He was quoted as calling for "one hundred skilled black revolutionaries who are ready to die." Another less riotous gathering was held at another church down the street, where the aunt of the boy who was shot was in tears and describing how the boy had defended himself against gunfire with the lid of a garbage can.

The next night there was a riot in Harlem. When accused of starting the riot, Jesse Gray testified that he had been, in his words, "trying to finish a document on European history." The city got a court order against any demonstrations by Gray above 110th Street. He held a rally at 109th Street. "It's the landlords who should be enjoined from operating in Harlem, not us," he said. He talked of fearing for his life. He quoted police officers as saying, "There's Jesse Gray, let's get him." After the riot, Mayor Wagner asked Martin Luther King to visit New York. King advised the city on civil rights. A few years later, the police department formed its first civilian review commission.

The rent strike marked Jesse Gray's heyday. After protesting with rats, he faded in and out of politics. He ran for mayor and dropped out after accusations that he had faked his petition signatures—a long list of names in alphabetical order, all allegedly witnessed by Jesse Gray. In 1970, he ran for Congress, arguing, "The money from five days in Vietnam could rebuild Harlem." His campaign slogan was "Nobody is behind Jesse Gray except the people." He lost. He won a state Assembly seat in 1972 and lost it in 1974. He was in the news and nearly sent to jail when his wife said he was not paying child support. He was accused of being a Communist, even though he probably wasn't. And he was accused of distributing anti-Semitic literature, which he did. His son, Jesse Junior was arrested on several occasions for selling drugs and finally sent to prison as a heroin dealer. Gray was quoted in 1977 as saying that he had not dropped out of the civil rights movement and that his silence should "not be misinterpreted as sleeping." Shortly thereafter, he went into a coma, which lasted for several years until his death, on January 2, 1982, in the Bronx. I don't know how he died, but I know that at the time tuberculosis, cancer, and diabetes were epidemic in Harlem. In 1990, the *New England Journal of Medicine* reported that men in Harlem had a shorter life expectancy than men in Bangladesh. A study in 2003, even after rapid gentrification had transformed parts of the neighborhood, indicated that one out of every four children in Harlem had asthma.

Once, I trailed off through Harlem, along streets I'd never been on before, looking for Jesse Gray's old headquarters or anyone who might remember him. I walked with a pace enthused by the knowledge of this rat-affiliated man, seeing the usual telltale signs of rats that I now see on all my walks everywhere in the city since spending time in the alley, but now also seeing the ghosts of rent strikers, of ancient community activists, of renters rising against rats. I walked down 125th Street, still the Main Street of Harlem, where some empty lots have evidence of rat infestation and some lots have brand-new national chain stores, and where one lot was surrounded by a wooden wall decorated with quotes from famous black Americans, such as Malcolm X: "Armed with the knowledge of our past, we can chart a course for our future. Only by knowing where we've been can we know where we are and look to where we want to go." Gray's rent strike headquarters had been demolished; like a lot of Harlem, the neighborhood is full of new housing now, some of it affordable for the people who live there, some not. I asked some people on the corner about Jesse Gray, an older couple. The man couldn't remember; the woman squinted her eyes. "I didn't know him personally, you see, but I understand that people spoke highly of him," she said. Then I saw a guy sitting on a beat-up old chair and leaning against a fence, right across from where Gray's old building had been. He was talking to himself and I wasn't sure if I should bother him, but I took a chance and said hello. As it happened, the guy remembered Jesse Gray and he remembered Jesse Gray's old building—he'd been in the neighborhood for decades. "Man, there's a lot of history in that spot right there, I'll tell you, that's for sure," he said.

*Chapter 8*

# FOOD

THE DIET OF the city rat is garbage, the refuse of man. But which garbage? Which particular kind of refuse? And exactly how much trash does a rat eat? One evening toward the end of summer, my thoughts traipsed in these repulsive particulars. It seemed to me to be a perfect night for a rat's food-gathering. If a wide-winged hawk at dawn might be satisfied with a steady wind and a clear sky as he ascends over a river basin made flat and treeless by the ancient run of a river, then so might a rat be pleased upon arising at ten o'clock on a Thursday night in the alley: the tide of garbage was just coming in, the doors of the restaurants opening for the jettisoning of garbage bags, then slamming tight, like clams. In the streets, there was light summer-evening traffic—young people darting into bars, older people coming more slowly out of bars, a man wandering alone, walking, stopping, walking a little more. I took my position at the front of the alley, just in from the evening street.

I heard the sounds of nighttime in an alley: the far-off moan of hydraulics from a truck digesting trash; the toss of ice and stale flower water from a delicatessen into a gutter; a garbage bag, first lofted and, soon, crashing to the ground with a *splatch*. Exterminators often note that rats are attentive not only to the sight and smell of trash but to the very sounds it makes, and my observations confirmed this. Initially, I supposed that the rats were waiting in their nests for the garbage's arrival. But then, sometime later, after watching the next fresh garbage bags thrown into the alley, I saw the rats come to the food again, and I realized that they were already out of their nests, already in the alley, roaming, foraging through bits of the previous evening's garbage, licking up fetid water, a portion of the two ounces they require each day.

And then they began to eat. Immediately, I thought of S. A. Barnett's 1956 rat eating study, summarized thusly by Jackson in 1982: "Mice tend to be nibblers; rats are more gluttonous." An example: a healthy male moved out from his nest area on the east side of the alley. He proceeded quickly along a curb, paused at a gap in the curb, then raced across the open space. He paused again, then moved back behind the garbage bags. This was a characteristic rat movement in the alley, a targeted series of bursts and pauses. Unfortunately, I could not see behind the plastic bag, so I do not know if the rat made his own hole—a simple rip—or if there was a preexisting hole, but quickly the rat was inside the garbage bag. The reader may ask why I did not go behind the bag and investigate more thoroughly, and my response is twofold. First, I was careful not to disturb the rats, for I was there to meticulously observe but not disrupt, and second, each time I approached the trash berms, even when I was fairly certain that they were ratless, an announcement would come over my internal public address system and, by way of reminder, say, "What the hell are you thinking? *There could still be rats in there!*" It was Coleridge who wrote, "Fear gives sudden instincts of skill."

A rat in a garbage bag is a keynote detail of the city landscape; if a rat were considered natural and people flocked to alleys to watch them gorge on the city's offscourings, the urban rejectamenta, then I would gladly send a postcard of my alley, which would, in such a world, be considered practically pristine, a wildlife refuge. Once in the bag, the rat is free to forage, using smell and touch and taste. The view from outside the bag is of the black plastic writhing and stretching. In my rat alley on this particular evening, the rat appeared to be working hard to consume one particular piece of garbage. I could not see the rat; I could see the rat lumps, though. Specifically, three short movements in the rat-shaped lump on the outside of the bag were followed by a larger expansion, at which point the lump moved to another area of the bag's interior. In four minutes, the rat emerged through a hole in the base of the bag with something in his mouth. On the western edge of the alley, adult rats were occupying other plastic bags, while younger rats were eating *around* the bags, tugging at scraps that fell from the holes gouged out by the larger rats. A few minutes later, with so many rats appearing, I was thinking of bears at an Alaskan salmon stream.

In the space of half an hour, it became clear that among the five garbage bags on the one side of the alley, the rats preferred the food in one garbage bag over the others. They were taking bits of this food back to their nests, so apparently satisfied were they with the safety of the alley and the quality of the food. The food that they were eating was white and stringy, and I had no idea what it was.

It is written in the rat literature that a rat would starve in an alley surrounded by raw vegetables. Pioneering studies of rats eating garbage were conducted in an alley by a New Yorker named Martin W. Schein, who was born in Brooklyn in 1925. During World War II, Schein fought under General George Patton, in the Battle of the Bulge, after which he returned to New York to work as a rat catcher for the city. He went to Baltimore to catch and study rats in the alleys alongside Dave Davis—Schein was once named Baltimore's honorary rat catcher. When Schein died, in 1998, a memorial said: "Imagine Hemingway in looks and in his direct, nonsentimental style, yet generous and kind to a fault." Schein founded the Animal Behavior Society in 1964, and after he finished studying rats he went on to work with turkeys. An experiment of Schein's that is today considered a classic in the field of animal behavior is called the "turkey on a stick," which showed that all that was necessary to inspire a male turkey to begin the mating-related behaviors was the female head. When working with garbage, Schein went into a block in Baltimore and dumped out a lot of garbage cans into one spot and separated it into edible and nonedible garbage. "In urban communities where natural rat-foods have been restricted by construction of buildings and the paving of streets, the rat has become dependent upon domestic refuse as a source of food," he wrote in a paper entitled "A Preliminary Analysis of Garbage as Food for the

Norway Rat." (Foods that would have been considered "natural" would include such things as plants and insects and small animals.) Schein found that on average one third of the garbage was edible. His hope was one day to be able to predict the number of rats in an area from pounds of refuse, but as best I can tell, he moved on to his turkey studies before accomplishing this. He had already shown a positive correlation between the number of rats and the amount of garbage. After analyzing the edibility of Baltimore garbage, he collected a lot of garbage from three different places—a college cafeteria, a grocery store, and a freight terminal—and he started feeding it to a group of rats. Schein had trapped rats from various alleys in Baltimore and transferred them to cages in an unused barn: city rats caged in the country. Here is a list of foods that the rats ate, in the order of rat preference:

| Garbage Rats Liked, with the Most-Liked Garbage First | Garbage Rats Didn't Like As Much, Least-Liked Foods First |
|---|---|
| Scrambled eggs | Raw beets |
| Macaroni and cheese | Peaches |
| Cooked corn kernels | Raw celery |
| Cooked potatoes | Cooked cauliflower |
| Cooked oatmeal | Grapefruit |
| Cooked sweet potatoes | Raw cauliflower |
| White bread | Raw potato peels |
| Raw corn on the cob | Raw carrots |
| Raw beef with bones | Green peppers |
| Raw sweet potatoes | Raw cabbage |
| Raw beef, no bones | Radishes (with tops) |
| Corned beef hash | Cooked spinach |
| Fried chicken | Plums |
| Bananas | Cooked cabbage |
| Cooked carrots | Apples |

I looked at this list frequently while ratting and found it to be a good guide to seeking out rat feeding points in the city in general and in my rat alley in particular. Of course, the list is not completely reflective of modern New York City garbage; it does not mention fish garbage, which might be a bigger part of the rat's garbage diet where I was, so close to the Fulton Fish Market. (At Peck Slip, I once saw the carcass of an Atlantic salmon that appeared to have been chewed upon by a rat, though the chewer could have been something else, I suppose.) Still, the list proved quite accurate. For instance, the rats in my alley rarely touched the raw carrots that were often scattered around, while they seemed to love cheesy Italian dishes. Schein noted that rats may have a preference for sweets and an aversion to spicy foods, and I would only add that, while I am not disagreeing with him on this point, an exterminator who was based in a Puerto Rican neighborhood in East Harlem told me that the rats there have learned to enjoy spicy garbage. This exterminator hypothesized that rats grow to enjoy the ethnic foods of the ethnic group in whose neighborhood they live. Post-Schein rat food studies corroborate this observation to some extent; this rat adaption is more technically described as a "local food dialect."

Likewise, while apples are listed here as unfavoured by rats, I once watched a rat work hard to keep an apple. One night, just inside the fence surrounding City Hall, I spotted a rat who had found an apple core; the rat was standing in the tall green grass of the park outside the building. I approached carefully, but somehow the rat noticed me, and it immediately began running north along the concrete base of the fence. Then, seeming to sense that I was following him—jogging alongside him, actually—the rat jumped off the concrete and back into the grass, looking almost pastoral as it galloped through the tall, summer-breeze-swept green. About twenty-five yards later, he jumped to the sidewalk and ran along a concrete barrier used to block access to a construction project, then down into a sewer. The apple would not fit through the grate on the sewer. The rat pushed, trying to make it fit. Then, as I approached, the rat, startled, jumped into the sewer and reached up to pull at the apple, again to no avail. Fortunately, I had my night-vision monocular in my backpack; I took it out and was able to get close enough to the sewer to look into it and see the rat looking at the apple. As I crouched down and gazed into the sewer, an Israeli couple who were tourists and didn't speak much English wanted to know where I got what they thought was a video camera but was actually the night-vision monocular. I tried to explain to them that I was just watching a rat in the sewer. At the time, I was with my friend Matt, the poet, and he and I pointed to the rat—if you knew what you were looking for, you could see the rat's head somewhat, even without night-vision gear. The couple nodded a lot, but in retrospect I don't think they were understanding me. I am not certain that rat watching translates, anyway.

OH, TO PONDER THE DIGESTIVE systems of the city, to consider the vast and mundane civic processes through which the city rat is nourished and the alley filled—for in the alley I can see the city as organism itself, a creature that consumes in unimaginable quantity, that excretes, eliminates, expels!

On that late-summer evening, I continued to watch the rats emerge from inside the black plastic garbage bags in my rat alley, and I still could not identify the food that so appealed to them that they continued to carry off from the bag. I moved closer, and at first the rats stopped, seemingly aware of my presence. I stood very still. After a few minutes, when they returned to eating, I walked up the alley again, this time along the wall. I was closer now, maybe twenty feet from where they were carting away the garbage, which was stringy and white, almost gooey. Then the back door to the Irish restaurant opened. The rats scattered and froze. I froze. A man threw out garbage, but did not see me, as far as I could tell. I waited, my heart beating loudly. The door closed. I continued to keep still. In a minute the rats returned, to pull again at the substance. Meanwhile, some young men and women came out of a bar and into the alley, to light cigarettes and talk. The men were dressed in jeans and T-shirts, and the women's beach attire seemed to clash with the alley's aesthetic. Again, the rats scattered, though this time they returned *before* the bargoers left. Now, the rats had hit a rich vein of the stuff they were eating—they were shuttling it back and forth, unbeknownst to the bargoers.

Finally, I realized I could go around the corner and check the restaurant's menu. After so doing, I went back into the alley, where the rats had begun fighting, and they were fighting over this garbage—two rats, screeching, attacking. One rat ran. I could see the other with his paws toward his mouth, eating the white substance, which I now saw clearly was one of the daily specials, chicken potpie.

## Chapter 9

# FIGHTS

THE CITY IS a place to which people migrate. It is a place where, as was (and still is) the case with rats, citizens-to-be arrived on ships in huge numbers, such that they swarmed ashore, finding homes in hovels and shanties, along dark streets and alleys, taking their place at the very bottom of the social hierarchy, near the settlement's bowels. In the nineteenth century, twenty-five million people came to America through New York. Millions of Germans arrived to escape war, and as New York was becoming America's largest and wealthiest city, the Irish, fleeing their great famine, came looking for work, looking for food. In the 1830s, one thousand people climbed out of the bottom of ships anchored offshore every day, and aside from living in pits, the newly arrived immigrants socialized. For many years, one of the things they did to relax was cram into little saloons, sometimes called sporting men's clubs, and stand around dirt pits and watch rats fight.

The most renowned rat pit was a place down in the seaport called Sportsman's Hall, owned and operated by a sportsman and rat fight impresario named Christopher Keybourn, better known as Kit Burns. Kit Burns was stout and red-faced, portly but muscular. He wore muttonchop whiskers. When he was dressing up, he wore a bright red shirt and suspenders. As more and more immigrants arrived in New York in the mid-nineteenth century, the upper classes tended to look at the working classes as "bestial," as *Harper's Weekly* put it—with rough hands, rude demeanor, tanned skin, and ragged clothes. People said Kit looked like a rat-fighting dog.

Kit Burns was born in Donegal, Ireland, and he came to New York as a boy, in a huge wave of Irish immigration—two hundred thousand Irish arrived in New York City around 1830. As a youth, Kit played with the dogs at Yankee Sullivan's Sawdust House, then the most famous dog-fighting parlor in lower Manhattan. Kit opened up Sportsman's Hall in 1840, at 273 Water Street, in a neighborhood described by nonresidents as "a slum of moral putrefaction." Kit's neighbor was a dance hall owned by John Allen, also known as "the wickedest man in New York." As a rat fight impresario, Kit made enough money to bring his parents over from Ireland and then his brother, who became a policeman. Kit made his own alcohol, which he sold in his saloon. He also drank his own alcohol exclusively and considered it a sign of the success of a strict twenty-glass-a-day drinking regimen that, when he fell down a flight of stairs, he was up and about in a matter of days. He was associated with a gang called the Dead Rabbits, a Irish working-class gang that defended the neighborhood from anti-immigrant nativist gangs like the Bowery Boys. Among these men, he was sometimes referred to as a "rodentary magnate."

Upon entering Sportsman's Hall, the rat fight patron first passed through the saloon, which was decorated with pictures of boxers and lithographs of hunting scenes and of people camping in the woods. Two of Kit's favorite dogs hung stuffed over the bar. Jack was a black and tan that had once killed one hundred rats in six minutes and forty seconds, an American record; Hunky was a dog-fighting champion who had died after his last victory. The bar itself was said to hold 250 decent people and 400 indecent ones. The rat pit was just beyond the bar. It was a wooden-walled oval on the dirt floor, seventeen feet long, eight and a half feet wide, with benches and boxes for the patrons. The rats entered in a wire cage the size of a large pail; they came in fifty at a time, rats screaming and hissing. When the dogs saw the rats released, they howled, setting the rats into a frenzy. "They galloped about the walls in different directions, meeting and crowding into a file in one of the corners, where they tried ineffectually to scale the top of the pit," a rat fight attendee wrote. "Then they would separate again and run frightened about the floor, trying every crevice and corner. One or two ran up the trousers and legs of the cage-holder, whence he composedly and carelessly shook them again." Jocko the Wonder Dog, a London-based rat fighting dog, was said to hold the world's record, having killed one hundred rats in five minutes and twenty-eight seconds.

Sometimes, Kit featured ferrets or weasels killing rats, but rat killing without dogs was considered a slower sport, more suited for women and children. On rare occasions, men fought the rats. A New York correspondent covering a Philadelphia rat fight described one such scene: "Then came a horrible spectacle. Quick as lightning the man plunged his hand into the mass of rats, seized one by the back and carried it to his mouth—with a squeak and a crunch, the lifeless carcass was tossed aside with a broken neck." When men fought rats, the man was expected to bite the rat's head off. This often resulted in the man's face being bloodied from rat bites. Even Kit was disgusted by this—he was said to have thrown a man out of his place for trying it. And yet when Kit died, Kit's daughter married a rat killer, Richard Toner, alias Dick the Rat.

The rats themselves came from the alleys around the docks. Jack Jennings was the brother of Harry Jennings, another rat pit owner, and Jack used to set out at night with two large canvas bags, a piece of iron wire, a crowbar, a jackknife, a trap that caged the rats alive, a lantern, and a large bottle of what he described as oil of rhodium, which he claimed kept the rats from biting him. A reporter followed him on a fall evening in 1866. Jennings went into alleys downtown and a stable on Front Street, near the Seaport. In the stable he turned on his lanterns and spotted some rats. Then he set his trap, and while he waited, he rubbed oil of rhodium on his hands and arms; he lay down on the ground and, crawling toward the rats, grabbed one after the other, quickly "giving them the sack," as he described it. He sold rats to the rat pits at fifteen cents a rat. The rats that the rat pits did not buy would often be sold to make gloves. Jack Jennings averaged 150 rats a night. Kit Burns used to say that the man who caught rats for him had "a gift," and that his methods were secret. "Lots of folks have tried to find it out," Kit said, "but 't ain't no use. It'll al'us be a secret."

A RAT, AS WE KNOW, lives in a colony, traversing a discrete range, and so did Kit Burns. And in his own neighborhood, among the boatloads of people immigrating to New York, Kit Burns was well-known and well-liked—full of rat fight stories. Outside of downtown, people were not as fond of him, especially Henry Bergh, the founder of the Society for the Prevention of Cruelty to Animals.

Physically, Bergh was the opposite of Kit Burns. Here is a description from *Scribner's Monthly Magazine:* "Nature gave him an absolute patent on every feature and manner of his personality. His commanding stature of six feet is magnified by his erect and dignified bearing. A silk hat with straight rim covers with primness the severity of his presence. A dark brown or dark blue frock overcoat encases his broad shouldered and spare, yet sinewy, figure." Bergh lived uptown, on Fifth Avenue, and he had a country house in the Hudson River Valley. He was the son of a shipbuilding magnate who had built warships for the government during the War of 1812. Originally, Bergh had hoped to be a writer. He wrote stories, poems, and plays; a piece called "Human Chattels" satirized the trend of wealthy New York mothers attempting to marry their daughters to European royalty, and "A Decided Scamp" was a comedy that few people thought was funny. In London, the reviews of one of his poems dismayed him. "Look at that!" Bergh said to his publisher. "They have literally skinned me alive." In 1862, he was made secretary of legation at Saint Petersburg, Russia. He wore a uniform, and when he did, he was surprised to see people kowtow to his rank. He began attacking animal cruelty. "At last I've found a way to utilize my gold lace," he said. When he returned to New York, in 1866, he founded the Society for the Prevention of Cruelty to Animals. Soon, people recognized him in the streets. He was known as "the ubiquitous and humane biped." "That's the man who is kind to the dumb animals," people on the street would say.

Bergh prowled the city for mistreated animals. "On the crowded streets, he walks with a slow, slightly swinging pace peculiar to himself," *Scribner's* wrote. "Apparently preoccupied, he is yet observant of everything about him and mechanically notes the condition from head to hoof of every passing horse." Bergh stopped carriage drivers to inspect the horses. If he deemed a horse lame, then it would be sent off in a horse ambulance, a contraption that Bergh had developed. (He also invented animal drinking fountains.) If a horse was suffering, then Bergh would have it put down: today, officers of the ASPCA drive around the city in what look like police cars, and they still carry guns, a remnant of the time when they might shoot a horse in the streets if it was suffering. If Bergh felt a cow's udder was too full of milk while a cow was walking to market, he would then force the farmers to milk their cows on the spot. He often gave impromptu speeches in the street, and if horse drivers or anyone else complained about treating an animal humanely, then Bergh would raise his walking stick or throw the men to the ground. Bergh convinced upper-class marksmen to shoot glass balls instead of live pigeons and exposed the cruel and unsanitary conditions in which milk cows were kept in basements beneath breweries and fed distillery garbage—the swill milk crime, as it was known. During the 1860s, Bergh turned his attention to dog fighting, which Kit Burns also dabbled in, and then to rat fighting, the area of entertainment in which Kit Burns reigned supreme. Bergh's campaign was effective, and by 1867 the *Evening Telegram* wrote that ratting has been "put down" by that "irrepressible suppressor of cruelty to animals, Mr. Bergh, and now it no longer delights the assembled throngs of Battery Roughs and Bowery Boys." In a letter to an associate, Bergh wrote, "[O]ne of our chief achievements is, *the breaking up of all the leading Pits.*"

The only pit still operating was Kit Burns's Sportsman's Hall.

KIT WAS USED TO RAIDS; he was wary of traps, of anyone new in his club. He'd engineered the exit in the back of his bar like an escape tunnel, a narrow corridor designed so that the police could be blocked by one or two men while the sporting men escaped out the back. But now Bergh was onto him. A policeman came down through the skylight one night, an overhead raid; he caught Kit's patrons in the middle of a rat fight. The men were all brought into court, where the judge complained about the smell of the men from Kit's neighborhood. Eventually they were released. But the raids continued. Finally, Kit was hauled into jail. In court, Kit hired a well-known defense attorney, William F. Howe of Hummel and Howe, who argued, first, that the men did not set the dogs against each other; and later, that the men were only rat fighting. Howe conjectured that if rat fighting was made illegal, then before long oysters would be banned, oysters being the most popular food in New York, until a little while after 1878, the year pollution closed the last oyster beds. Also, Howe made a big show of the pain that an oyster would feel when the oyster was chucked with a knife between its shells; he argued that if the people discontinued ratting, then a man might one day be arrested for consuming oysters.

The judge added, "Only if he chews!"

At this remark, the portion of the courtroom filled with Kit Burns's neighbors broke into hysterics; the portion filled with members of Bergh's entourage later complained to the judge that this remark was not funny.

Kit's own defense of rat fighting was based on his view of the rat as a nonanimal, or non-anything even. "Mr. Bergh calls a rat an animal!" Kit said. "Now, everybody of any sense knows that a rat is a *vermin*. Bergh takes up for the rat and won't let us kill rats because he thinks they're animals. Wouldn't he kill a rat if he found one in his cupboard? Of course he would. But, would he kill a horse if he found one in his yard, or even in his parlor? Of course he *wouldn't*. Why? Because a horse is *an animal,* but a rat ain't. I *know rats*. I know they're vermin, and they *ought* to be killed. And if we can get a little sport out of their killing, so much the better."

Henry Bergh continued to hunt down Kit Burns, leading raid after raid on Sportsman's Hall. Meanwhile, all around the Water Street rat pit, religious reformers were taking over the saloons and dance halls for prayer meetings. John Allen, who was known as the Wickedest Man in New York, rented his dance hall out for prayer meetings, and Kit had many offers to do the same, all of which he initially declined. Then he wrote an open letter to Henry Bergh in the *Herald,* inviting him to come and speak on rat killing at Sportsman's Hall:

> Since Johnny Allen closed his crib I've been thinking of myself and my business. I own and train lots of dogs and I kill any number of rats in the pit. A good many first rate fellows, a little rough in speech, may be, but trumps at heart, call in to see the dogs kill the rats, and I've some that are just a little touch above anything in town in that line. You see I trained them pups myself. But what I was going to say is this:—I like everybody to have a fair chance, and I believe Mr. Bergh's on the square from the word go! Now, if that gentleman will just call on me any day, I'll fix it so he can talk to the crowd, deliver his lectures to them—and there are sometimes two or three hundred present—and if he can show us that we are cruel and are doing wrong, why all I've got to say is, I'll burst the pit and give away or sell out my dogs; and I've some of the finest ratters out. I feel kind of dubious about this thing; and yet I can hardly make myself believe it's wrong or cruel to kill rats, however it may be about fighting dogs. But I'd like to talk to Mr. Bergh about these things, and I'm sincere, if he'll only call on me, at my place, No. 273 Water Street.
>
> Respectfully,
> Kit Burns

Bergh did not respond to Kit Burns's letter.

Seeing that his neighborhood was going uphill, Kit decided to try renting his place to the religious leaders. Soon, a prayer service was held in his saloon every day for an hour beginning at noon.

"Do you intend to give up your business, Mr. Burns?" a prayer leader asked him, as he waited anxiously near the bar, which—during the prayer services, anyway—was not serving Kit's homemade alcohol.

"Not much," Kit said. "Not if I know myself. No, gentlemen, the games of the house will go on the same as ever. As soon as those 'fellers' leave, we're going to have a rat-killin'—a bully time—and all the fun you want."

The leaders of the prayer service attempted to tell him that the money would eat at his conscience. "Oh, they can't come that over me. I'm too old for that," Kit said.

In December of 1869, Kit's favorite dog, Belcher, died in a fight with a dog from Brooklyn. Kit said that in retrospect he'd thought the dog had been a little off since the prayer meetings. "He was never exactly himself after it," Kit said. "It wasn't so much the praying as the singing that took hold of him." The fight in which Belcher died was to be the last dog fight in Sportsman's Hall. Kit subsequently rented the entire building out for three years. It became a mission and home for wayward women, called The Kit Burns Mission. For a short time, Kit opened a smaller saloon down the street, called The Band-Box.

At last, Henry Bergh got wind of what would be Kit Burns's last rat fight, on November 21, 1870.

THREE HUNDRED RATS WILL BE GIVEN AWAY, FREE OF CHARGE,
FOR GENTLEMEN TO TRY THEIR DOGS WITH
COME ONE, COME ALL!
THERE WILL BE A GOOD NIGHTS SPORT AND NO HUMBUG.
TICKETS 25 CENTS

By eight o'clock four or five dozen dead rats were in a pile and one dog was still going. Bergh, who had been informed of the fight, snuck in and, by some accounts, carried a lantern under his coat. When the police raided, the sportsmen shouted, *"Douse the glim!"* Bergh was ready with a lantern, foiling their escape. Their claim that they were merely preparing for a boxing match wasn't convincing enough to prevent thirty-nine men from being arrested. The dogs were taken away, the ones that were fighting destroyed. The surviving rats were thrown in their cage in the East River. *The Herald* said that Bergh had finally got Kit and his men: "He will make them squall worse than the unfortunate rats which were dumped into the East River by the police." Kit was upbeat upon arrest, but reportedly depressed soon after. He was sued for the value of dogs that the police had destroyed. His family moved to Brooklyn, and Mrs. Burns, in the *Sun,* invited Henry Bergh to visit her there "provided," as she said, "the gentleman will have the kindness to bring his coffin with him." Though he had survived a knife in his neck at Kerrigan's saloon a few years before, Kit caught a cold and died, before trial.

The sporting men were all acquitted. A man testified that it was not a dog fight but a rat fight, which was still considered less reprehensible. "The blood they tell about were rat blood, that's wot sort of blood it were," the sporting man said. And the judge suggested that if they stopped dogs from killing rats, they would next have to make it illegal for cats to kill mice. There was a funeral for Kit a few days before the trial. A parade followed his body from his home in Brooklyn, all the way to Calvary Cemetery in Queens. "The excitement in the neighborhood was most intense, and crowds gathered around the house for some time previous to the hour set for the funeral," the *Herald* wrote. "The crowds poured into the place and gazed in the face of the dead with as much apparent reverence as if the deceased were a high-toned, honorable, moral and religious light in the community."

A few tributes to Kit Burns were published in the papers, such as this one, which is itself a tribute to a rodentary man: "Departed from this life yesterday . . . one whose birth was humble, and who did not aspire to congress, a member of no religious denomination; a life-long enemy of the police magistrates; a Fellow of the Metropolitan Society of the Slums; a professor of the art of dodging the penitentiary; an enthusiastic believer in 'Rings'; the subject of much pious objurgation, and the hero of many a newspaper sensation; the beloved of our 'ruling classes,' and the pet of the 'Water Street Warblers'; a genius in disguise; a Democrat by birth, and a 'Dead Rabbit' by association was the dear departed, which is his name, as it shall be forever written on the hearts of New Yorkers, is Kit Burns."

Most newspapers wrote that they were happy that Kit Burns had died. "We are glad of it," wrote the *Citizens and Round Table*. Henry Bergh was among those well pleased. "I drove him out of New York and into his grave," Bergh said, a few years later.

In addition to preventing an incalculable number of animal tortures, Henry Bergh went on to found the Society for the Prevention of Cruelty to Children. Thankfully, he did eventually stop rat fights and the like, though such events persisted for some time—people naturally want to gather in crowds and eat and drink and cheer and sometimes get into brawls. Some historians argue that the end of rat fights did not come until the next inexpensive crowd-pleasing sporting event was finally embraced by the growing number of inner-city residents in New York and all over America: baseball.

## *Chapter 10*

# GARBAGE

A LATE-SUMMER EVENING, almost fall. A delicious evening that was cool but not cold, an evening of haze-blurred stars beneath which, as I crossed the Brooklyn Bridge, the East River fought the rising tide to rush into the harbor, the bay, the ocean. And then I was down from the bird's-eye view of the bridge and into the twinkling buildings, then the side streets, then the alley, which is not really light and not really dark but locked in its semi-sickly fluorescence, in the side-street twilight that is its Arctic White Night. The rats were out, grazing peacefully in the two garbage berms, in the Chinese and the Irish trash. So many rats, at least a dozen visible now, some large, some small. Rat life in the alley can be a harried blur, a wash of rodentary citizenry, or it can be a kaleidoscope of the recognizable, a rat-a-tat-tat of what almost immediately becomes a harmonized familiar—and so I consider these questions: Are there more? Is this colony growing? Or am I just noticing rats that I did not notice before?

I watched as the rats entered and exited in what was now their familiar pattern: holding tight to the wall and, leaving their nest area, taking a first few tentative steps, then halting, then making that exhilarating rat charge, the wall to one side, the open alley to their left. Another halt and then another charge, and in a moment a rat was up on the crinkly black of the garbage bag or under the bag or through the well-chewed hole and into the semi-eaten spoiled and not-so-spoiled once human food. Each rat takes a particular course that is nearly the same and yet slightly different from the next rat. In so doing, they were carrying food back, though some also ate the food in place. I once read a rat study that suggested that the likelihood that a rat will eat depends on how safe he feels at a food acquisition site in relation to how safe he feels in his nest, which, it has occurred to me, is not unlike a human apartment dweller's consideration when ordering takeout.

Boldly, I stepped out into the alley, as if stepping from behind my blind. I was more confident by now, more at ease with the feeding patterns of the rats, though still a little jumpy. And when I stepped out, when I walked up into the alley, the rats initially hesitated at the sound of my footsteps, but then as I moved slowly, easily, they seemed to take less mind. They stayed on their course, making their jerky flights along the walls, into the bags, answering the call of the discarded-by-humans food. Precisely how many rats were they? I couldn't say at this point. And before the reader scoffs at this remark, I suggest they go to Grand Central, to Penn Station, to a Grateful Dead concert on a farm in the Oregon country, or whatever passes for a Grateful Dead concert, now that the Dead are gone, and try to count—it is not for nothing that the masses are called the masses.

It's considerably more difficult to distinguish rats *en scène* than it might seem to the armchair rat watcher—this is something that I was understanding more and more. I will, however, say that I could see between eight and ten rats at any given moment, which may not sound like a lot, but those rats appeared to be part of a larger relay team, one group replacing the next. Also, I lost count as to how many rats were in the heaps of garbage bags at a given time. The bags were animated now, each bag churning, heaving—a bar brawl in a pup tent.

As an experiment, I stomped. The three rats I could see moving at that moment froze. After the count of just four seconds, they started up again. Likewise, the garbage bags were quiet, then they resumed their soft rustle. The old papers of Dave Davis show that determining the total number of rats from visual counts is somewhat unreliable; the most reliable method is to trap the rats, tag them, release them back into the city, and keep trapping until a total count is reached. Using my own more unreliable visual methods, I would estimate between fifty and sixty rats lived here. But implementing a more general rule sometimes used by rat professionals—if you see one, then there may be ten in the vicinity—I estimated about one hundred rats living in the alley, hidden away in holes, basements, underground vaults. And to think that the first time I looked down the alley all I saw was a dead end! Still standing midalley, I now sensed multiple dartings, fast blurs in the corner of my eye, each causing me to momentarily consider an alley evacuation. I stood still for a little longer, nonetheless, attempting to focus again, to calm my nerves, to concentrate on the energy of these seemingly caffeinated quiverings—to become an all-sensing, outer-focused, night-vision eye. How bold these smallest strokes of nature!

I stood proudly at the top of the alley, looking down the incline toward Fulton Street, standing in a place that I felt was out of rats' way, seeing the waves of pedestrians passing back on nonalley land and feeling only slightly repulsed, and a wave of calm finally came over me—until, looking left, I noticed that a small crack in the sidewalk was moving and then noticed that it was a rat. One rat, two rats, and then a brief rat squabble as the first rat attempts to return through the hole and crosses paths with a third rat coming up. It was a new rat source, another nest entirely, which I had only accidentally discovered in my moment of haughty self-congratulation. I turned the corner and looked down Edens Alley toward Gold Street and saw more rats coming up from holes in the street, through gaps in the cobblestones: rats hoisting themselves up, ramming their snouts up from below the street, pulling out their front legs and then heaving, hauling themselves up to quickly find the edge of the shiny curb, the trace of a wall, to scurry, to scatter.

This was a lot of rats for one person to handle. I got out of there. I walked across Fulton Street, back to the little traffic triangle where I could still see the garbage in the alley and watch the rats with binoculars and wonder what else I was missing. I set up my little portable camping stool, and it was then that I looked down at the ground and saw an engraving in the stone and slowly realized that the name cut into the little square I was standing in was nearly synonymous with garbage. I realized that I was standing in John DeLury

Plaza. As rats and refuse are concerned, it was as if I had suddenly realized I was standing on a nose on Mount Rushmore, or in the basement of Monticello.

As Neptune controlled the seas and all its creatures, so for many years did John DeLury control the city's garbage flow—the sanitation workers were his union mermen, his trust-bound dolphins. John DeLury was the first and longtime leader of the Uniformed Sanitationmen's Association (U.S.A.), and when I left the rat alley that evening and began digging in books, I learned that John DeLury was the man who changed the term *garbageman* to *sanitation worker.* I also learned about the very first time that his workers stopped working, a time when the city filled to the brim with trash—a time that would surely be written about by rats if rats could write. "We may deliver garbage, but we are not garbage," was something John DeLury was fond of saying. Men and women hanging off the backs of city garbage trucks still remember John DeLury today because John DeLury went to jail because of garbage, in an act of sanitary disobedience.

DeLury went to jail in 1968 when John Lindsay was mayor and Nelson Rockefeller was governor. Lindsay was young, good-looking, and Yale-educated, a Republican from the city's Silk Stocking District on the Upper East Side, a district so named in 1897 because of the colonies of wealthy residents along Park and Fifth Avenues. Lindsay was nationally prominent as a leader who was interested in giving minority neighborhoods more power and as the mayor of "Fun City," though the city was not so fun at that moment, since the mayor was borrowing heavily to finance a huge budget deficit as well as feuding with the governor, whom the city needed to bail it out. Nelson Rockefeller was a nationally prominent leader who was a trustee of the Museum of Modern Art, founder of the Museum of Primitive Art, soon-to-be-vice-president, under Gerald Ford, and the first governor of New York to set up an office in New York City, where he governed except during legislative sessions. Also, he was a Rockefeller. John DeLury was not that young—sixty-three—and he was not considered especially good-looking—he was white-haired, bespectacled, short, tough-seeming, and always gnawing on his pipe—and he was not a Rockefeller. DeLury was born in Brooklyn, the second of thirteen children. He quit high school in 1921 to help support his family and he worked for fifteen years on Wall Street, until he found a job working at a dump that increased his salary from $100 a month to $135. He organized his fellow dump workers—building many small garbage-related unions into the U.S.A. In 1938, when John DeLury first became president of the U.S.A., the salary for a forty-eight-hour workweek was $1,800 a year. By the mid-1960s, the sanitation men earned $6,424 to $7,956 a year. DeLury won the union a forty-hour workweek in 1956, twenty years after a forty-hour workweek had been made a federal law. In the seventies before DeLury retired, the salary of a sanitation worker was equal to the salary of policemen and firemen. Many New Yorkers hated John DeLury for this very reason, but the sanitation workers loved him.

DeLury was known to shout and yell and scream at public officials, but somehow they still considered him reasonable, a navigable storm. He had an acute sense of precisely how far he could push City Hall with his demands. "He's a bulldog. He can wear anybody down, never lets go, no doubt about it. But he's not an unreasonable guy," a sanitation department commissioner once said. Before the strike of 1968, the sanitation workers' union had only resorted to one short strike, in 1960—DeLury always preferred to negotiate. The secret of his success was said to be the collection of file cards describing the life details of each of the ten thousand union members and their families, all of which he kept in the basement of the union hall. "Only God can guarantee one hundred percent of the vote," DeLury would say. People who weren't DeLury noted that DeLury came very close. One of the last leaders of Tammany Hall, the old New York political club, once said, "I would today rather have John DeLury's sanitation men with me in an election than half the party headquarters in town." DeLury's men respected him too. "He had a lot of involvement with the members," a union member who remembered him told me. "He was very honest. He spoke our language, and everybody knew him as John. He used to say, 'I'm your leader. I need you.' He wasn't above anybody. But he was a tough son of a bitch too. You wouldn't want to have him as a father-in-law." After work, DeLury would sit in his Greenwich Village apartment and listen to classical music, reading biographies and labor histories.

On February 2, 1968, seven months after the sanitation workers' contract had expired, John DeLury called for a mass meeting in City Hall Park, the ancient commons of the city, once called the Fields. It was a cold, gray day. The sanitation workers arrived at seven in the morning, filling the little patch of green and then spilling out into Broadway and Park Row. DeLury climbed on top of a car and took the pipe out of his mouth. He shouted out the terms offered by the city officials. The crowd rejected the offer. They called for a strike. DeLury didn't want the strike, but over and over the crowd shouted, "Go, go, go!" So DeLury called a strike. The courts ordered the workers back to work later that day, but they didn't go. Garbage immediately started piling up. Mayor Lindsay asked the city's highway workers to collect garbage. They declined. Three days later, DeLury asked the striking workers to collect garbage at hospitals and schools. They did; signs on their trucks said, "We're taking it away without pay." The courts called the strike illegal. As a crowd of striking sanitation workers cheered, DeLury surrendered himself to the city sheriff and walked into the Thirty-third Street jail.

Garbage, garbage *everywhere*—ten thousand tons of garbage piling up every day instead of heading off on barges, instead of being dumped on landfills. The streets of New York looked as if they had been covered with snow and plowed except that the huge snowbanks along the streets were made of trash. In some neighborhoods—in Harlem and East Harlem, for example—it didn't look as if the snow had even been plowed; the trash was inches deep on the streets and the sidewalks. For this, New Yorkers despised John DeLury and his men. One judge said of the strike, "It's blackmail, it's extortion." A nationally syndicated columnist pointed out that the union was Mob-tainted—a deputy commissioner of sanitation had been killed gangland style. *Life* magazine conceded that the laws governing municipal workers' unions were "archaic and senselessly rigid." But most New Yorkers were, as the *Daily News* editorialized, "fed up with kicks in the kisser from strike-happy public employee unions."

As if on cue, as if he knew that by doing so he was playing his trump card, the city health commissioner invoked the specter of rats, which Paul O'Dwyer, an attorney for the union, commented on after the strike was over: "[T]he rats, which had bitten four hundred slum-dwelling children last year, might indeed invade the middle-class and wealthier sections of our town. While we did not seem to get terribly excited while the vermin were attacking the children of the impoverished, our society did get itself in a state of white heat at the thought of rodent escalation."

Mayor Lindsay was furious. He declared a health emergency and demanded municipal employees transfer to sanitation jobs. The municipal employees refused; they would not break the strike. Now, there were rumors of a general strike, something that has never happened in New York and that would have completely paralyzed the city; the Teamsters were considering shutting down all trucking. Finally, Mayor Lindsay asked Governor Rockefeller to call out the National Guard to pick up garbage. Several editorial pages supported the idea; people reportedly telephoned the governor to induce him to accept the mayor's request. But the governor refused. He went on TV to say that if ten thousand soldiers picked up trash, the city would have half a million tons of it on the street at the end of two months, due to the soldiers' lack of experience in handling garbage. "We'd be buried in it, ladies and gentlemen," the governor said.

A guard brought DeLury from jail to the governor's office on Fifty-fifth Street for negotiations. An arbitration panel suggested a $425-a-year raise. DeLury exploded, left the room, returned, then accepted the offer. Mayor Lindsay rejected it. DeLury told the governor to go ahead and call out the National Guard. According to the union officials present, Rockefeller then recalled a strike at a company owned by his family, during which the National Guard had been called in and men and women were killed, at which point the governor supposedly said, "There must be another way," and then began to weep so that the labor leaders got embarrassed and looked at each other and left the room. It seemed as though the governor was being sensitive and sensible; on the other hand, DeLury had supported Rockefeller in the gubernatorial election and was mainly using the governor to go over the mayor's head. In fact, everybody was using everybody else; it was a bacchanalian feast of garbage-related favors.

When the governor collected himself and the labor leaders returned to the room, the governor authorized the state to take over the city's sanitation department, to pay the workers until a contract could be signed. At five in the morning, DeLury was escorted by the sheriff to the union headquarters, where the two hundred shop stewards had been waiting since ten o'clock the night before. They roared when DeLury walked into the big room. Like the governor, he got all choked up. By nine o'clock, the sanitation workers were out picking up trash. DeLury was back in jail.

THE GARBAGE WAS MOSTLY PICKED up in a week. DeLury was released after a few days. Over the next few weeks, DeLury, with the help of a public relations firm, set out to illuminate the sanitation workers' working conditions. He stood by poster-size photos of trucks that regularly caught on fire. He toured reporters through rooms heated by stoves fed by garbage found on the street. He demonstrated human-limb-eating hydraulics. He shouted about a hernia rate higher than in the logging industry (according to a professor from Springfield College in Massachusetts the union brought in to test the physical stresses of the job). He had the wives of amputees testify to the hardships involved given that sanitation workers didn't receive workers' compensation for any injury on the job. On TV, DeLury pointed to a grime-encrusted drinking fountain. "Now, fellas," DeLury said to reporters, "I know you gotta make do, and I know you gotta be on the sovereign right of the state and all that, but remember, we're people too."

When the settlement was reached weeks later, the union got a raise: $425 a year, bringing their maximum wage to $8,356 a year, almost as much as the wage of a city sewer worker at the time. The arbitration board also increased the union's pension fund, a trend in municipal union contract deals. A number of politicians and historians argue that giving the unions pay raises and pension increases was a mistake that nearly killed the city, financially speaking; people still gripe today about the union's power in the sixties. On the other hand, a few years after John DeLury settled his union's strike, when the city was completely bankrupt, when the banks would no longer give it any money to cover its debts, when the federal government wouldn't lend it any more money and the *Daily News* summed up President Gerald Ford's attitude toward New York and all cities with the front-page headline FORD TO CITY: DROP DEAD, when the city was a day away from total fiscal ruin, the municipal unions dipped into their pension funds to lend the city money—albeit reluctantly—and the city stayed alive. I like to think of this as a case of the rats *not* leaving the sinking ship.

After the garbage strike, Rockefeller, for his part, lost political support by not having called out the National Guard. Mayor Lindsay won reelection, though not with the support of John DeLury. "You know there's something about a jail," DeLury said on one radio show after he was released. "I've never been in a jail before. There's a rapport when you get there. You know when you have that common denominator that your freedom is taken away from you, you warm up to the other individual who is incarcerated."

"So you think Mayor Lindsay could be warmed up?" a reporter asked.

"I don't know about him," said DeLury.

LIKE A POND FILLED WITH RIPPLES, the city is filled with circles that overlap and intersect, that share a focus, or a deli. Sometimes I think the city is naturally conducive to coincidences in the same way that

Plains states like Nebraska and Oklahoma are conducive to twisters, in the same way that mountain lakes are conducive to lightning. From John DeLury Plaza, I could see the rats running in and out of the dark hole that was in the back of the alley, though I had still not determined exactly how deep the hole was or any of its dimensions, and it wasn't until a couple of days later that I realized the building that the hole was behind was the headquarters of the Uniformed Sanitationmen's Association—on Cliff Street, just around the corner. Naturally, I was blown away.

A couple of days later, I got up the nerve to knock on the door of U.S.A. headquarters. When the door opened, I met a guy who told me that rats used to run up and down his legs all the time when he was starting his sanitation career in the Bedford-Stuyvesant section of Brooklyn; that more people apply to be sanitation workers than police officers because the pay is better; and that, yes, he remembered John DeLury, his old union boss, who had died in 1980.

"He found no levity in any kind of sanitation jokes," the guy told me.

I began to chuckle but then just nodded—this guy was pretty serious himself. But he softened up when recalling John DeLury: "He told me something once and I'll never forget it. What he said was, 'When a union is a protection for incompetence, then you're losing sight of what a union is. When you represent a productive and hardworking workforce, unified, you're boundless.' I'm paraphrasing but that's what he said and I'll never forget that."

Rats seem to operate in waves of activity, and on that evening at what felt like the very beginnings of fall, I stood for a while in John DeLury Plaza and watched the rats climb in and out of the hole in the back of the alley—I was obviously even more interested in that hole now, knowing its geographic relationship to John DeLury and the U.S.A., but was still feeling a little nervous about it, as it was still dark and deep and rat-filled. And then, when the current wave of rat activity seemed to stop, I headed home.

I think that I am disgusted by rats as much as most people, and as I have tried to impart, I am naturally no rat-alley observer, no rat-secluded soul, but rather I might possibly enjoy having a drink at a bar at night on the way home from work more than anyone—the customary, social, and nonrat thing for a city dweller to do. So that sometimes, after I was done ratting, if it wasn't too late, I would stop in at the bar in my neighborhood and have a beer and watch the last innings of a baseball game while looking over my rat journal. That particular evening, I walked into the pub, which was crowded, and elbowed in at the bar. I was just standing there talking with my friend Dave, the artist, whom I kept up-to-date on my rat experiences, who was rat-interested like many New Yorkers and, of course, city dwellers of every kind. Dave, as it happened, was talking with his friend John.

I was excited about rats and ratting; as autumn was approaching, I felt as though I was really settling into my work; I thought I was, as far as rats went, finally beginning to see. With glass in hand, I extolled the alley and the rats to John, who was unusually interested in hearing about them, which is to say he was not immediately repulsed. Then I mentioned John DeLury. John looked at me in a kind of startled way. He stopped drinking. "*John DeLury was my grandfather!*" he said.

*Chapter 11*

# EXTERMINATORS

IN THE CITY, given the absence of rat-eating wild mammals and great numbers of birds of prey, the natural predator of the rat is the exterminator. Exterminators hang out in basements and crawl into old, dark spaces that no one else wants to crawl into. They attempt to think like rats and then kill them. Exterminators roam the city with traps and poisons, in marked and unmarked vehicles, depending on the customer's preference. They work in small mom-and-pop operations and sometimes with large firms. In some four-star restaurants and fancy hotels, working as an exterminator is like working as a spy behind enemy lines; if you are discovered, your existence will be denied. Once, I met an exterminator on the subway. He was dressed in overalls and carried boxes of tools and supplies on the kind of luggage cart used by someone repairing a computer or copying machine. It was the first day of winterlike temperatures at the time, and I asked him how the weather was affecting business, if the weather was driving rodents into people's homes. As he was getting off the subway at a downtown station, he smiled and nodded repeatedly. "Things are starting to happen," he said.

The first professional rat catcher in New York was the first professional rat catcher in America, Walter "Sure Pop" Isaacsen, who opened up a shop in Brooklyn, in 1857; he sold poison grains guaranteed strong enough to kill elephants and relied heavily on ferrets, which he raised on a farm in the countryside.⋆ In 1893, the

⋆ Traveling exterminators probably existed even earlier in America's cities, men selling their poison powders and extermination services from pushcarts. In

city's star rat catcher was Frederick Wegner, who arrived from Bavaria and made his name, first, by ridding Brooklyn's Prospect Park and then Greenwood Cemetery of rats. When there was a rat infestation in the Central Park Zoo—the rumor was that the elephants had been attacked by rats—he was immediately called in and caught 475 rats in his first week; he used traps because the zookeeper was worried about poison around the elephants. (Recently, in the National Zoological Park, in Washington, D.C., two red pandas, an endangered species, died after eating rat poison buried in their exhibit.) Harry Jennings was a well-known Manhattan-based exterminator with a shop in SoHo. When he died in 1891, an editorial writer said: "While not employing a very excited social position in the city, there were few more useful men than Harry Jennings." In 1936, the National Association of Exterminators and Fumigators voted to change the name of the profession from *exterminator* to *pest control operator.* William Buettner, a second-generation New York City exterminator, was the president of the association at the time. Exterminators were concerned that the word *exterminate* made for too great an expectation for the vermin-infested, exterminators as a group being, above all, realists. "Bill contended that the word *exterminate* suggested a permanency to a customer that was not possible to provide, while *control* more closely described the service provided," a pest control historian explained. An alternate choice was *verminologist.*

The city of New York founded the Rodent Control Unit in 1949. Optimistically, they still use the term *exterminator.* The number of

*The Ratcatcher's Child*, a history of the American pest control industry written by Robert Snetsinger, an early exterminator named Solomon Rose is reported to have set up a company in Cincinnati around 1860; he appears to have been selling antirat powders to Northern soldiers during the Civil War—at least until December of 1862, when Ulysses S. Grant ordered the expulsion of all Jews "as a class" from territory that is now in parts of Tennessee, Kentucky, and Mississippi. Abraham Lincoln rescinded the order in January of 1863. Snetsinger theorizes that Grant was attempting to get rid of the itinerant peddlers who followed the Union Army and sold them things like whiskey. "Grant's action as seen by Lincoln was unconstitutional and is viewed by modern Jews as evidence of Grant's anti-Semitic tendencies," he writes. Snetsinger also notes that German-Americans' expertise with the exterminating chemicals came from innovations in their homeland with brightly colored fabric dyes, which were all the rage in European fashion in the late 1800s.

exterminators in the bureau of rodent control fluctuates with the strength of the economy. For instance, when Juan Colon began exterminating in the early 1980s, when funding was available, he was one of ten men who exterminated from Ninety-sixth Street all the way north to the tip of Manhattan Island. Toward the end of his career, when funding was no longer available, he was the only one. Today, Colon is remembered as the only exterminator ever to tie a rodent to a rope and walk it back to the bureau office, eager to show off its impressive size. "I was young and crazy," Colon once said.

THE ALGONQUIN TRIBES CONSIDER THE hunter to be their best man, and if I were an Algonquin, I would see exterminators as the city's best men, so many of whom, while spending my time in Edens Alley, I have had the opportunity to meet. I can report that among exterminators, there are many different types. An example of a publicly supported rat catcher is Larry Adams, who is the highest-ranking exterminator working for the City of New York. Although he has worked in the private pest control industry, he has spent most of his career exterminating for the city; I think of him as the people's rat catcher. Larry works in Brooklyn, but he is likely to be called anywhere in the city. If you are in New York and you see a headline such as GIANT VAMPIRE RATS TERRORIZE LOWER EAST SIDE!—which was an actual headline a few summers ago—then you can be certain that Larry is on the job, baiting and filling up giant vampire rat holes. In the case of the giant vampire rats, even Larry was impressed not so much with the rats' size, which was merely typical, but with the number. "We had a tremendous kill over there," Larry said.

Larry is fifty-three years old and of medium height with a graying mustache and a slightly mischievous grin; he moves cautiously, occasionally interjecting quick bursts. When I stopped by the rodent control office in Bedford-Stuyvesant where he is based, he was wearing dark work pants, a blue work shirt, a walkie-talkie, and a flashlight clipped to his belt. He was born in Brooklyn, in the Bedford-Stuyvesant section, the son of a truck driver from North

Carolina. Growing up, he never saw a rat; his neighborhood was clean, and in his memory he sees his neighbors hosing down their stoops and sweeping, always sweeping. "I never saw a rat until I was a teenager," he said. Then, Larry's family moved to a housing project in Williamsburg, a section of Brooklyn that was filled with breweries up until the sixties. When he walked to school, he often passed rats, feeding on hops and barley. He worked for the sanitation department in high school. "I made thirty-two dollars and fifty-four cents one time," he said. "I remember that." He played football but then injured himself and sat on the sidelines as his team won the city championship. "I was a ball of confusion, so many issues I was dealing with," he recalled. He dropped out of high school and got a position at the health department. He was hired as a health department laborer, one man on a crew of people responsible for cleaning out abandoned lots and trash-plagued houses—a job he took just until he could get something better. Although it only paid $5,600 a year, he did not consider it a disgrace, especially during the city's fiscal crisis in the seventies. "People were like, 'You got a *job!* With the *city!* How did you ever get *that!*' " he said.

Larry had never seen so much garbage as he did working that job. "I remember saying that I never knew anyplace existed like this," he said. "Garbage as high as that. Garbage going to the second-floor window in the backyards. You could actually step out the window on the garbage and work your way down to the ground." This was in the seventies, when neighborhoods were burning out in cites all over America, when the once-clean streets in Bedford-Stuyvesant were decomposing, when the city seemed as if it were dying. One day Larry was out near Coney Island, clearing out the house of a mentally ill woman who had never thrown anything out. "The house was filled front-to-back, wall-to-wall, floor-to-ceiling with every type of garbage, junk, with pails of human feces, urine, everything," Larry recalled. "I got three skin infections working that particular job. There were flies and maggots, and rats. They used all the best workers out of all the crews to do that job, and I was doing my thing that day. I was coordinating stuff. I was working for everyone. And I didn't know it at the time, but the director was sitting about a half a block away, watching with binoculars. Before the day was over, a guy comes up to me and says, 'Hey, the boss, he's watching you and he likes your thing.'"

In a few weeks, Larry was a regional director in the department of health, working out of a rodent control office in Bedford-Stuyvesant, where Eubie Blake, the great ragtime piano player and songwriter, often poked his head in. "He'd stop by every day," Larry said. "He always had a thing to say. He'd make you laugh. He used to like to watch the girls on their way to the train station. He'd say, 'Now, you catch up with her and get her to hold on while I catch up to you.' It was like two separate worlds. In one world, Eubie Blake was this big famous guy and was a piece of American history, and in the other world he was just a neighborhood man I respected, an old man I listened to."

Larry went to school at night to study pest control, and soon he was moonlighting for a small pest control operator in Queens. He had just started working pest control when, on the night of his first wedding anniversary, he got an emergency termite call. "My wife had a fit," he remembered. "Boy she was *mad!* I got home at one-thirty in the morning after drilling and pumping chemicals. But I came home and spread all that money across the bed. I made a few hundred dollars, and in the 1970s two hundred dollars for one day's work was all right. So I laid it out on the bed and she said, 'Oh, *fine!*' We made up for it, I'll say. Everything was just fine." He also worked as a gypsy-cab driver and sold hats and other merchandise on the street. "Man I was nuts!" he remembered. "But I was married and we kind of adopted four of my wife's nephews and nieces—two of one sister's children and two of the other. It was something. There were good times, though, a lot of good times."

Larry hoped to become a city exterminator but there were no openings. Meanwhile, he was enjoying his work as a private exterminator. One Sunday morning, his wife sent him to a live-poultry market to buy a chicken. He spotted a rat running through the crowd of people in broad daylight. He immediately showed the owner of the market his card. "I'm gonna take care of this for you," he said.

"I mean, this guy would have sixty rats sitting on his fence in the morning when he was opening up!" Larry said. "And then across the street there was this semidemolished building, and a fish market down the corner dumped old fish in the building. I said, 'Jesus Christ! I gotta take care of *that* too!' But then a few days later, there were rats everywhere dead, just everywhere."

He worked on a crew of men who would become the great rat-killers of modern New York: Jimmy Clarke, who worked in Queens; Herbie Shackner, an exterminator in Brooklyn; Michael Castrulli, who worked on the Lower East Side. They were all at the famous Ann Street rat infestation, near Theatre Alley, where the woman was chased by rats to her car. Larry still remembers the Ann Street incident, when the woman was attacked by rats. "I've never seen so many rats in all my life," Larry recalled.

At last, he was promoted to exterminator, and now, after nearly three decades of experience hunting rats, he says he feels as if he has seen everything when it comes to rats, but he is still surprised on occasion. An example: "This was just two or three years ago, over in McGolrick Park in Greenpoint, and I had been out of work for four months—guy ran a light and tore me up. And anyway, I come back to work, first day, and I get a call from Rick Simeone, who's director of operations, and Rick, he said, 'Larry, you're not gonna believe this, but we got rats in trees.' I said, 'Get out of here! What are you *smoking?*' I said, 'Rats don't live in trees.' I said, 'Rats may be crossing over to use the limbs to get to a food source or something, but they don't *live* in trees, you understand?' He said, 'Larry, they're in the trees, Larry. There are *rats in the trees!*' I thought maybe he'd been drinking. So I said, 'Okay, we'll be over.' Then, I go over to where they're talking about. First thing I see is a rat sticking his head out of a hole in the tree. And then I look up. There's a rat on the limb. They were running all over the place in the daylight. So I pick up my walkie-talkie and I call Rick, and I said, 'Hey, Rick! *There's rats in the goddamn trees!*' "

With his expertise, Larry has developed his own rat-eradication techniques, such as concrete mixed with broken glass to keep the rats from gnawing through the concrete. "Sometimes, they'll still cut through before the concrete hardens. So sometimes, I use glass and industrial-strength steel wool and put it in with the concrete and make one big goop with it." He has also gained the respect of his friends. "They used to say, 'Look. Here comes rat man!' " Larry says. "Now, they are likely to say, 'Here comes Larry. Rats don't stand a chance with Larry around. Rats gonna get out of town.' "

If you hang around with Larry long enough, you realize that he sees the city in a way that most people don't—in layers. He sees the parks and the streets and then he sees the subways and the sewers and even the old tunnels underneath the sewers. He sees the city that is on the maps and the city that *was* on the maps—the city's past, the city of hidden speakeasies and ancient tunnels, the inklings of old streams and hills.

"People don't realize the subterranean conditions out there," he likes to say. "People don't realize the levels. People don't realize that we got things down there from the Revolution. A lot of people don't realize that there's just layers of settlers here, that things just get bricked off, covered up and all. They're not accessible to people, but they are to rats. And they have rats down there that have maybe never seen the surface. If they did, then they'd run people out. Like in the movies. You see, we only see the tail end of it. And we only see the weak rats, the ones that get forced out to look for food."

One day, I followed Larry Adams as he went out exterminating. His office is across the street from a vast housing project in Bedford-Stuyvesant. The basement is filled with old wire rat cages from the time years ago when the city used to catch rats alive in cages and sometimes put them on a Greyhound bus up to Albany for testing. There are usually one or two rodent control vans parked out in front of the place. On that cool, gray morning, Larry got in one of the vans with two assistant exterminators—Ralph Saunders and David Simpkins. Larry trained Ralph when he was starting out in rodent control, and Simpkins's father was a city exterminator. "I've got it in my genes," Simpkins said. A yellow light was flashing on the dashboard of the van.

"It's only for emergencies, really," Larry said. "Like now. We're going to a rat bite."

The rat bite was in the Bushwick section of Brooklyn, in an apartment building on the fourth floor. Bushwick is full of new immigrants to New York and America, the group who have, it might be argued, historically suffered the brunt of the city's rat infestations, not to mention other problems. At the rat-bite scene, the van met Gary Gaynor, a health department inspector, in front of the building. Gary was dressed in a yellow windbreaker and carrying a clipboard and an L.L. Bean backpack embroidered with the words RODENT CONTROL. "Sometimes you go into places, and you don't want to go in but you have to," Gary said. "I've been down in places where I know they're there, but I can't see them, and then I look up and they're walking around in the ceiling." The *they* he was talking about was rats.

The sidewalks were narrow and, in many places, crumbling; the street was lightly parked with old cars. Larry inspected the sidewalk for holes and rat-size basement entryways. He pointed to one hole in the sidewalk, a cragged circle of gnarled cement: "See that? They'll gnaw through concrete." He pointed down and then David put on a respirator and pumped a poisonous tracking powder into the hole. David used a giant bike-pump-like device that the exterminators refer to as the Bazooka.

On the way upstairs, the exterminators attempted to exterminate in other apartments. One door was padlocked shut. "There could be something in there but we can't get in right now," Gary said.

Upstairs, in the fourth-floor apartment that had called in the rat bite, a young girl was sitting on a couch, watching TV alongside her brother and her aunt. They were quiet as they watched the exterminators work. The woman held the girl, who seemed scared and clung tightly to her aunt. The woman showed the exterminators around the apartment, which was sparsely furnished but very clean. She said they were originally from Ecuador. When they talked about the rat bite, the woman sat down next to the girl and held her. Gary spoke kindly with them. Larry, meanwhile, looked behind the stove, where he found rat holes. Larry asked David to bait the holes with poison. In the living room, the girl remained quiet. "She got bit by a rat," the aunt said. The aunt did not seem to speak much English, but she managed to explain to Gary that the girl had been to a hospital, that she had been examined and was okay. The aunt then showed the exterminators how she and her sister block the back room of the three-room apartment with couch cushions each evening in hopes of keeping out the rats.

In a little while, Larry went down in the basement—it was dark and low-ceilinged. He spotted the place where the rats had been coming into the building and made a note to tell the landlord to patch up the holes.

Rat bites mean rat infestations, of course, and so Larry headed out into the streets to continue hunting—he had been working in this neighborhood recently. Down the street, Larry began knocking on doors.

"Health department," Larry said. "We're inspecting for rats and baiting."

"Good," said a woman, standing on her stoop.

Next door, a man from Trinidad enthusiastically invited the exterminators into his studio apartment. It was filled with Indian religious paintings and spiritual music CDs. The man said his girlfriend was upset by the rats in his apartment and no longer visited him. "My girl, she get scared," he said. He showed the exterminators a photograph of his girlfriend, and he held a loaf of sandwich bread by its plastic packaging's twisted end—the bottom of the loaf was chewed away. "You see that? The rat ate that," the man said.

Back on the street, a landlord named Joe Fragala approached the exterminators. He was wearing a Yankees T-shirt and jeans, and when he found out that they were exterminators, he thanked them. "The rat situation's gotten a lot better over here," he said, "maybe because you guys are baiting with poison." He pointed a few blocks away. "But over on there on Melrose, forget about it! They're going *crazy!* I've seen them, believe me."

Meanwhile, Larry had picked up on something. He was walking back and forth in front of a house and looking down intently. He seemed to be sniffing. Rat droppings were everywhere and he could smell them. It was incredible. He directed David and Ralph to stuff poison in several holes, and he followed David along, poking the poison down into the hole with an old broom handle that he carried with him.

Finally, he stopped before an old wooden tenement building. He inspected rat droppings and pointed out one dropping in particular. "Man, that one is two inches long. I don't know *what* made that," he said.

He knocked on the door of the building, and an old woman answered in a well-worn nightgown, which she cinched at her waist as she shooed Larry away. "No rats!" she said. "No, no rats!" The woman closed the door.

Larry walked down the stoop and then stood before the building for a minute. "They're in there. I can taste 'em."

He walked back up the steps and knocked again. This time the old woman relented. Larry went straight for the basement, where I could finally smell the rat smell he'd been talking about, which was now almost unbearable for him. As he walked down the tight, old stairs, he said, "Whew! Do you smell 'em? It's like a zoo down here."

On the floor of the basement were more rat droppings, huge and fresh. David and Ralph went to work. Meanwhile, Larry was walking around, looking down and scanning the floor and the walls—until all of a sudden, when he stopped and looked up toward the low ceiling. Something darted down, raced to the floor, where it stopped and looked right at him. It was a big rat. Larry smiled.

* * *

Another, very different kind of exterminator is Barry Beck. If Larry Adams is the people's exterminator, then Barry is the private pest control operator writ large, the highest form of the mom-and-pop exterminator, the representative of rodent control for profit. He makes a big point of not being an exterminator—it is a selling point, in fact. "We don't consider ourselves exterminators," he explained the first time I dropped by his office. "We consider ourselves pest control managers."

He gave as an example his work with pigeons, a species that, he stresses, he does not poison.

"Do we kill pigeons?" he asked rhetorically. "No. We *exclude* them."

Barry runs a big pest control operation—in fact, it is the biggest firm in the city. "There's two hundred and fifty pest control operators in the five boroughs," he says. "Two hundred and twenty of them are small. The other thirty are medium-sized and then there's us." He works on the largest buildings and is currently involved with the largest construction project in the city. Like a good exterminator, he doesn't name names, and in terms of rats, as well as mice, he has an unending market. "There's no city like this city. I mean, all cities have their problems, but in terms of pest control, this city is *so* infested that it's incredible. It's incredible!" Barry is married and has two children, a young son and a teenage daughter. He works constantly. "I'm a capitalist," he likes to say. "I've devoted as much time as I can to my business without getting a divorce."

Barry Beck is of medium height and medium build, and he works in an office in the middle of Manhattan, where, in a dark business suit, white shirt, and dark ties, he blends in with the non-rat-exterminating business community. Aside from being an expert in pest control, Barry is a salesman, and he began his career in sales out in the suburbs on Long Island, working alongside his father in the car business. "I rented cars, fixed cars, sold cars, did all kinds of stuff," he said. "I'm a car guy."

Barry's route to rodent control went like this: he went to college, got married, went looking for a job, answered an ad for a position with a cleaning company in the city. They wanted to hire him as a salesman in their pest control division. "I turned off, and I basically said I never saw a rodent in my life," he remembered. "I really didn't have too much interest, but they asked me to think about it and they called me and asked me to come in for a second interview, and the guy sold me a bill of goods. And he was right. The opportunities were unlimited for somebody who knew how to sell and knew operations. I moved up. I broke every sales record they ever had, and this was the largest company in New York at the time. I moved up the corporate ladder, rose to executive vice president. And then things were changing but nobody knew why, but my division was doing unbelievable. They were thinking of cutting my company car, which was the wrong thing to say to a car guy. I had a Toyota Supra Turbo—I always had the best cars. I said, '*Really?* You should offer me a chauffeur and a limo.' So then I called up Bob and Ted at Pioneer Pest Control, and I said, 'I'm looking for a job.' And they said, 'Okay,' and Bob said to me, 'Barry, I respect your boss at the company you're at now, and I would never call you, but you called me.' And then he said, 'What do you want?' We had dinner at a Chinese restaurant on Fifty-sixth Street between Lex and Third. That was ten years ago. Now, I drive a Mercedes convertible."

Today, Barry commands the largest pest control force in the city. "In the industry, the standard is, how many men do you have?" he says. "We now have a hundred." And he has worked tenaciously for his company to grow, constantly selling, constantly finding new clients, eating away at the competition. "When I started at Pioneer, they probably had twenty-five men," he recalls. "I went from working for the largest company to this rinky-dink company. My office used to have a view of Madison Square Park. I had a couch, the whole bit. And then I had a desk outside the chemical room in the hallway. I didn't care. This was a established business that needed to grow."

He has opened up new pest control divisions—his bird division, for example, which includes the pigeon exclusion department. To emphasize the deleterious effects of pigeon excrement to clients, he cites an incident a few years ago in which pigeon droppings cut through a cable on the Brooklyn Bridge, due to its acidity. The cable snapped and decapitated a man. "The object is to build out pests without pesticides," he says. "That's the future." He foresees a day when he will be hired to analyze a building's weaknesses, vis-à-vis pests and rodents. He sees this as a more humane pest control method, and humane systems are currently in vogue. People often ask that mice be trapped live, and once in a while people ask about live-trapping rats. "Every year I get somebody who calls up and asks us to trap the rats, and I say, 'Well, fine, but where would you like us to release them?' And the people always say Central Park." Barry's daughter is a vegetarian, and once she asked about the possibility of more humane ways of eliminating roaches. "I said if they come up with a way, I'll be happy to sell it," Barry said.

Barry can't believe how often people attempt to skimp on pest control. "They design buildings to support pigeons and for infiltration by rodents because they don't think about it. Grand Central Station, right? They just renovated it, right? Who knows what they spent on that, right? You know how much they spent on pest control? You know how much they budgeted? Nothing. I did all the extra work there, but they had to pay us out of the emergency budget."

Barry guards his client list carefully. "I got lots of clients, which I cannot identify," he said. "They're *big*. You wouldn't believe how big."

Once, he had the World Trade Center as an account. "I always wanted it, I could never get it," he said. "I tried like hell to get it. It goes out to the lowest bid to the littlest guy who does nothing. I can't compete." When the building was first bombed, in 1993, office workers evacuated but left food that they were eating all over the building. As a result, there was a rat problem. Beck was called in. "They were concerned, and they remembered who we were," he said. "We did top to bottom and they loved us, but after the emergency was over, then they had to go out to bid again. I'm not interested in a big-volume account with no profit. I'm a capitalist. If I can't make a profit, I don't want you. I don't need prestige. I got prestige."

As the biggest pest control firm in the city, Barry's company is often called in for the big jobs that smaller outfits might not be able to handle. "A couple of weeks ago, we had a call for a job," he told me once. "I'm not gonna say where, but it was a big job. The guys had to wear the protective suits. There were *inches* of droppings. This is in a place right here in Manhattan—Midtown. We go in there and we see carcasses and skulls, and we're thinking, *What's going on here?* And sure enough, we found out later that these people were eating them. Eating the rats!"

He leaned back in his chair, imagining these people eating rats, on purpose. "I'll tell you what," he said. "It's bizarre and incredible, but that is their culture. It's not my culture but it's their culture. People treat rats different in different places. There's a country in Africa where they worship the rat. It's true. I saw it on TV."

Barry's company, meanwhile, takes over lots of smaller companies. For instance, when I mentioned a smaller one that I'd heard of, he said, "I took over that company one and a half years ago—swallowed them up." Recently, he took over the company he started out with. "It was bittersweet," he said, "but we've been the biggest in New York City for about seven years. You just keep growing and growing and growing, and then the next thing you know, you're a giant. But you just keep working. It's like, if I have this huge sale today, what about tomorrow?" Despite all his success in preventing or even eliminating rodents, his respect for rats only grows. "They're survivors. They know how to live. They know how to keep going."

At his desk one day, Barry was all pest business. Before him were sample rodent traps and repelling devices that were being tested and a plaque that said, "Best wishes—National Cleaning Contractors." Inventing is a hobby for Barry. He invented the Extend-A-Wand, a compressed-air-sprayer extender. He also invented a kind of rodent bait simplification device, which is patented under the name Myrna-Baiter, named after his mother-in-law, Myrna. "I have a bunch of other ideas, but I have no time for 'em," he said between calls.

Each time the phone rang, Barry said his name forcefully and quickly and bluntly.

"Barry Beck," he said once, for example, and as he did, he quickly reached for a yellow legal pad.

"Yes, thanks for returning the call," he went on. "I heard you were interested in our bird eradication techniques."

Pause.

"Yes. What do *you* feel like, tomorrow?"

Pause.

"Sure, tomorrow at ten. *Great!* Would you like me to bring by some of our exclusion devices so I can show you some of our techniques?"

Pause.

"Okay, great!"

At this point, Barry hung up the phone and then pointed his finger at me and smiled. "I've been looking up for two and a half years at the pigeon dung on that building and leaving flyers every time I go by."

And then another call—from a rat expert whom he had been trying to reach to discuss a new rat exclusion device that a company had asked Beck for help in testing. The device is designed to scare rodents away with sound, though Barry is not sure it works. The secretary puts the expert through.

"Rich, it's Barry Beck with Pioneer Exterminating in Manhattan. How are you?"

Pause.

"Yeah, look," Barry said. "I was interested in getting a study done and was looking for a ballpark on a new ultrasonic on rodents."

Pause.

"Well, basically, everything I've seen about their unit is that they're at the initial stages. Basically, from an engineering perspective I think they might be on the ball, but from a rodent perspective they're in the Dark Ages. It's two guys and one of the guys has got a brother who's an astronaut—it's a big name, I forget. Well, I'm looking for a price on a study. My criteria would be, number one, are rodents expelled with the device? And, number two, if you put up barriers, how does the sound affect the rodent with partition walls?"

Pause.

"Okay, thanks." Barry put down the phone. "You know, there was a guy I really wanted to get to help us, but he's busy. I mean, this expert is good, but there is someone in this country that knows even more than him about rodents. That's Bobby Corrigan. He's backed up over two years to accept a consulting job."

I asked Barry if there would be any way for me to meet Bobby Corrigan.

"I think it's impossible," Barry said. "But if you could, that'd be something. He has done more studies and understands more about rats literally than anybody in the world."

*Chapter 12*

# EXCELLENT

SO I RAMBLED westward, away from my alley, off to the middle of America, because who wouldn't want to meet America's authority on rats and rodents, if they were me? Who wouldn't temporarily suspend observations in their rat-infested alley for just a short time so as to attempt to make contact with a man who knows more about rats than anybody else in the world?

Shortly after I first heard about Bobby Corrigan, I learned from an exterminator that he was going to be speaking at a barbecue out on Long Island thrown by a large pest control equipment supplier, in a town called Hicksville. Accordingly, I took a train out to Hicksville and caught a cab in the rain and, in the back of an industrial park, found the pesticide distributing company. There was a grill with hot dogs and hamburgers set up under a canopy outside, and people ate inside the warehouse, which was full of traps and poison. I listened to several presentations about mice and rats, and also about roaches, termites, and flies, and I met people in the pest control industry who were offering special barbecue sale prices on roach paste and fly traps and rodent poison. I sat next to a guy who heckled a speaker when she said she was going to run up to the front of the crowd as quickly as a mouse. "Oh, like you're going to run eleven miles an hour!" he said, to the laughter of the pest-control-operator-filled audience. Early on at the barbecue, I realized that Bobby Corrigan was not actually going to be there. Fortunately, I heard about a huge upcoming rodent control conference in Chicago. Not only was Bobby Corrigan organizing this conference in Chicago—he was scheduled to give the keynote speech.

So I took *another* trip, figuring that not only would I meet Bobby Corrigan but I might get an opportunity to see rats elsewhere, to compare and contrast them with my rats. I went to Penn Station and took an overnight train to Chicago, arriving at rush hour, at which point I was able to stop in the middle of Union Station and lean back against a wall and watch people as they streamed in and out of train-track exits and entrances, in and out of the exits to Chicago's streets, of the entrance and exits to a restaurant also marked with signs indicating areas for ordering food to go versus to stay. I smelled the food. I grabbed some. After that I pushed haltingly through the crowd and got on another train to Milwaukee—a city that seemed to me to be worth visiting before I met Bobby Corrigan as it has a long tradition of being ahead of the curve nationally in terms of pest control. None other than Dave Davis, America's first and greatest rat expert, once said, "I have visited a number of cities, and Milwaukee is surprisingly good—no, let's say spectacularly good—as far as rat control is concerned." In addition, I'd heard that the mayor was going to hold a press conference—perfectly timed, as far as I was concerned—regarding rat control.

My time spent in Milwaukee was hurried—I mostly ran around. If I had been observing my own movements through the city that day, I would have noted myself

- arriving at the station, then immediately seeing, right where I stepped down off the train, a rat bait station, a Protecta model, manufactured by Bell Labs, in which the bait was stale and partially eaten—a sign;
- wandering tentatively across town and passing by the Wisconsin Workers Memorial, a park that featured a sculpture that is also a time line noting the years 1911, when Wisconsin enacted the nation's first workers' compensation law, and 1920, when Wisconsin drafted the first unemployment compensation law; a

quote on a plaque read, "The 19th Century belief that unemployment was a matter of individual bad luck or bad character was deeply ingrained in Wisconsin and American culture, and the realization that in fact it was an unavoidable feature of the modern industrial economy came only slowly";

- checking into a great old hotel, the Pfister, a hotel from the days before Milwaukee suffered rioting and a huge recession and lost sixty thousand manufacturing jobs, from the days when Milwaukee was a brewery capital full of German refugees fleeing the revolutions of 1848 who knew how to make great beer and, due in part to their experience as chemists, rodent poison;
- wishing I could slip up to the Pfister's antique-looking bar and order a beer and wait for someone to ask me what I was doing in town so I could say, "Rats," but instead going up to my room and calling my wife, who told me to hurry up and call Don Schaewe;
- calling Don Schaewe, a rodent control official in Milwaukee, who was jealous that I was going to see Bobby Corrigan and suggested I come over and get an idea of the rodent control strategy in town prior to the Milwaukee mayor's rat press conference the next day, and who gave me directions to the Nuisance Control Field Office, which I subsequently gave to a taxi driver, who, when he heard the address, grimaced such that I asked him if it was a bad neighborhood or something, to which he replied, "Bad? *Bad* is not the right word. *Worst!* You cannot walk in the streets. Crime, drugs! I never pick up fares over there, not in my life";
- getting out of the cab, thinking the neighborhood didn't seem so bad, and then quickly knocking on the locked door and entering a nice office full of very nice pest control people;
- walking around the office as Schaewe showed off the large closet containing various rodenticides, which, he suddenly realized to his momentary astonishment, had a small mouse infestation problem ("Huh," Don said, "that's like somebody breaking into a concentration camp"), then listening as Schaewe talked about,

first, the old days when the rodent control staff used a meat-based rat poison and ground the meat on an old meat grinder that is still there, and second, a recent innovation in which computers are linking crime to rodent infestation;

- taking a ride with Don to the neighborhood where the mayor was going to hold a rat press conference the next day, and seeing evidence of serious rat infestation in the alleys behind the neighborhood of modest one- and two-family homes that either looked as if they had seen tough times but were well kept or looked the same but weren't well kept and had lots of trash littered on their lawns—along with evidence of serious infestation that included rat trails, rat burrows, and dead rats;
- getting a ride from Don back downtown, where he dropped me off at a sausage place (where I ate some of the best sausage I've ever tasted), from where I proceeded to the Milwaukee Public Library, where I looked at beautiful Audubon prints and read about Milwaukee's rat control history, noting, for example, the numerous "Starve a Rat" campaigns, the free rat films the city showed in rat-infested neighborhoods (titles included *Listen to the Rat Man, Professor Rat,* and *The Rat King*), and the guy who went around in a rat suit giving kids rat pins—the health department's uniformed sanitarian in the fifties, Lieutenant Archibald Kowald, used to say, "We look upon people as souls, and we want them to have a better world to live in";
- then heading to Maeders, an old German restaurant that Don had recommended ("The last time we went there my wife wanted to try duck so she asked for duck and they gave her a whole duck!"), where I drank two huge glasses of German beer and had a sampler plate that included goulash, schnitzel, pork loin, and what was billed as "Germany's favorite soup," a soup made with duck and liver pâté, not to mention dessert;
- then—after I'd paid the bill and read the thank-you letters from overweight, dead celebrities on the wall, in addition to thank-you letters from healthier-looking celebrities like Boris Karloff

("If I had a whole regiment with me I could have done justice to the meal you placed before me . . .," wrote Karloff, in a pleasing cursive)—heading into the alleys of Milwaukee looking for rats and having trouble: feeling fat, bloated, practically rolling through the alleys downtown, looking for trash and rats and seeing neither, realizing that the alleys I went into were incredibly clean, with the trash neatly stored, with only a moderate amount of grease, understanding, at last, that I was too disgustingly sated to really have my heart in ratting;

- stumbling into the hotel room where I took some aspirin and collapsed.

THE NEXT DAY, A CABDRIVER reluctantly took me to the mayor's rat control press conference in the rat-infested neighborhood that I'd been to the day before. Television news crews were in an alley by the time I arrived; they were waiting to interview the mayor, John Norquist. They were willing to interview the mayor about rats, though I soon learned that they were undoubtedly interested in the sexual harassment case he was involved in at that time. He had recently admitted to having an affair with a woman on his staff. The woman had charged that the mayor had threatened to withhold block grants to Hispanic and African-American neighborhoods if this woman, who is Hispanic, "refused to heed Mayor Norquist's sexual desires," as one news report put it. Myself, I was merely interested in talking to the mayor about rat control.

While we all waited for the mayor to arrive, Don Schaewe was pointing out all the rat holes in the area and the disparity between trash-filled lawns and cleaned-up lawns for the TV reporters, and then talking to one of the veteran exterminators, Gabriel Perez. Perez reminisced about the time he was working in a neighborhood and a woman invited him in her house to see a rat. She showed him the room with the rat in it. He entered, looked at the rat, and then, before he knew it, the woman locked the door behind him. "She wouldn't let me out until I killed the rat, so I had to kill the rat," he recalled. He also recalled the extensive training that he'd gone through years before, when there was considerably more government funding for rodent control. "We just talked rats eight hours a day," he said. "Sometimes I would leave and my head would be just like *rats, rats, rats*."

In a while, a police car with its lights on pulled up to the back alley and then a shiny black SUV pulled up behind it. The mayor got out of the SUV. He was tall and charismatic with a warm grin and looked a little like a lithe John Wayne in a business suit. When the conference began, the department of neighborhood services showed off some charts; officials spoke about the rat crackdown in the neighborhood: the plan was to go door-to-door with flyers that alerted people to the neighborhood's "chronic and increasing rat population." The flyers instructed people to clean up their garbage or be fined. Then the mayor spoke:

"Rats eat human food. That's what they eat. So if you don't want rats around, don't feed them. That's the message we're trying to get out." The mayor held up a chart and smiled for the camera. "If people don't like rats, don't feed them. It's as simple as that. People should look in the mirror first."

The mayor turned to the cameras. "Any questions?"

There were no questions.

At this point, the mayor and a couple of dozen volunteers from community organizations began to walk down the streets to hand out flyers. The mayor was all smiles. His bodyguard—a big African-American guy, an ex-cop, in a suit—was a few steps behind him, wearing no expression at all. The mayor stopped at the first house on the corner, but it looked as if it were abandoned—until a big, hungry-looking German shepherd came out on the roof of the second floor and began barking ferociously and then a little girl appeared on the front porch.

"Why aren't these children in school?" Rosa Cameron, an alderwoman from the area, said. The alderwoman was among the community group members. She bent over and looked at the girl's skin and thought she might have worms. "This is one of the worst blocks, as far as my district goes," Cameron said. She stood up and began walking again. "You know, anytime you push for the poor, it's a fight."

The mayor, meanwhile, continued to greet the residents. He lingered at the home of a man and a woman whose home was relatively immaculate; he knocked on the doors of people who rented the old, worn-out houses, and the renters were either incredulous or wary. One man came out his front door and continued talking on his phone and watching the mayor as the mayor approached, smiling and sticking out his hand. The man kept his phone to his ear when he shook hands with the mayor. "Hey, Mayor," the man said, "you just come back at night and they'll be running at your feet."

The mayor proceeded to the next house, walking along jauntily, smiling, waving. At the next house, the man and the woman at the door stared at him. "Hi," the mayor said. "Did you get the word?"

"Yeah, well, they're foreclosing in a couple of days. I'm gonna be out," said the man.

"Okay, well, wherever you move, don't feed the rats," the mayor said, before smiling and waving good-bye.

On the way up the front walk to one house, the mayor's bodyguard ran up to the mayor and whispered in his ear just as he was about to knock on the door. The alderwoman noticed something was up. "I don't know what he's going in there for. His bodyguard's getting nervous," she said. After the bodyguard whispered to him, the mayor quickly turned away from the house—it seemed as if the mayor's bodyguard thought he had steered the mayor away from someone who was dangerous.

The mayor visited a few more houses, and in a few minutes the event started to break up.

I was ready to go. Don Schaewe wished me well. "Say hello to Bobby Corrigan for me when you get to Chicago," he said.

I was trying to figure out how I was ever going to find a cab when one of the mayor's aides offered me a ride back to the hotel. I got in the shiny black SUV. The mayor was in the front seat, leaning back, his long legs barely folded in. He appeared relaxed and was making me feel comfortable: hearing that I was from New York City, he said that his brother-in-law, a musician, had played with Tito Puente, the salsa player from the Bronx. The bodyguard drove and he was doing a good job of being quiet and formidable-looking. As we all drove out of the rat-infested neighborhood and into the beautifully renovated downtown, the mayor was going over some of his impressive credentials as an advocate for urban renewal and job creation; he talked about some of the factories he had encouraged to open up in the area; he talked about job creation plans. I was making notes when I made a remark that questioned whether crime was not somehow linked to poverty.

"If you're looking for a poverty angle—well, if people *really* wanted to get a job they could," the mayor said, and turned around in his seat to look at me. "Do you know why people commit crime?" he asked rhetorically. "I mean, it's not like it's *La Bohème* and Mimi's in the back room starving. It's fun. It's a thrill. You break into the house, and that's more fun than dipping metal in a chemical in a heat-transfer plant. Instead of working, you go break into a house. It's more fun."

The mayor laughed and turned back to look out the front window of the SUV.

"Until you get caught," he went on.

"Right?" He turned to his bodyguard. "Ask *him,*" he said, motioning to his bodyguard. "He was a cop."

"It's a thrill," the bodyguard said. I looked in the rearview mirror and saw the bodyguard looking at me and smiling.

A few months later, Mayor Norquist settled the sexual harassment charges out of court. He announced that he would not run for a fifth term. Also, Rosa Cameron, the alderwoman I'd been walking around with, pleaded guilty to funneling $28,000 worth of federal grants intended for community groups into her campaign fund; she was sentenced to jail time and testified against other city officials.

After I finally got to Chicago and eventually home, I called Don

Schaewe and he told me that the rat population in the neighborhood had gone down for a while. He was as optimistic as he could be: "That's a tough neighborhood. You're always gonna have some rats there. We knocked 'em down, but you'll never get rid of them."

RAT CATCHERS OF AMERICA! MICE trappers of all the United States! Men (mostly, though a few women) representing places such as Smithereen Pest Management Services in Evanston, Illinois; and Western Exterminator Co., in Anaheim; California, and Wil-Kil Pest Control in Sun Prairie, Wisconsin. People from Varment Guard in Columbus, Ohio, and National Bugmobiles in Victoria, Texas, who, when they get the frenzied phone call, lay out the poisons that the rats eat, inspiring the rats' deaths! Academics who study rats and work with the pest control industry or with public health institutions! Those who attempt to keep this land rodent-free—they would all be at the Rodent Management Summit sponsored by *Pest Control Technology* magazine and held at the Courtyard Marriott in downtown Chicago and orchestrated by Bobby Corrigan! I was in rat expert heaven. The publisher of *Pest Control Technology,* who opened the conference, described the event as a gathering of, in his words, "the finest minds in the pest control industry," all gathered in Chicago, that city of which the poet Carl Sandburg has sung:

> Come and show me another city with lifted head singing so proud
> to be alive and coarse and strong and cunning . . .

I would think of living there, in Chicago, singing with lifted head, except that the city I know is New York—my birth city and, as a result, the only city in which I feel comfortable with lifted head singing. But while there I was in heaven, bathed in the resplendent light of rodent knowledge, meeting, for instance, William Jackson.

Jackson is America's rodent expert emeritus. He got his start working on the first wild-rat studies with Dave Davis in Baltimore, as I have already mentioned. With Davis, Jackson went out into alleys and studied, among other things, cat feces, noting a low amount of rat parts in cat feces, thereby indicating that alley cats and alley rats generally keep their distance. He has helped governments all over the world with their rat problems. When I met Professor Jackson, at the cocktail party on the second night, we talked about the rats that he'd investigated in the Enewetak Atoll in the Marshall Islands in the 1960s—the rats that had survived nuclear testing in the Pacific. He said that the rats had survived the blast by staying deep down in their burrows, and that, upon investigating, the only abnormality that he could find was a change in the structure of the rats' upper jaws, a change that did not seem to hinder the rat in any way. During his presentation, Jackson lectured on rat poisons. He talked about the time after World War II when America was especially chemical-happy—antibiotics had just been used successfully on the battlefield, and DDT had broken a wartime typhus outbreak in Italy. "The attitude there was that chemistry was going to save us, that chemistry was going to take care of all our problems," Jackson said. Warfarin, the first modern anti-coagulant rat poison, had been discovered accidentally at the University of Wisconsin at Madison, in 1948, when a chemist noticed that cattle died of internal bleeding after eating spoiled sweet clover. The "clover chemical" was soon isolated and fed to laboratory rats, in an effort to find a cure for arthritis. The rats died.

After Jackson described the invention of warfarin, the pest control operators seemed as if they were getting a little bored—my feeling was that they mostly just wanted to know how to get rats. But Jackson pressed on, explaining that Norway rats became resistant to rat poisons in the seventies. The resistance, first noted on a farm in Scotland in 1960, was soon spotted on farms in America and around the country. In 1976, 65 percent of the rats trapped in three neighborhoods in Chicago survived warfarin, and in Wisconsin, there was a report of rats subsisting on warfarin-treated grain. Also in 1976, the Bureau of Pest Control found that 12 percent of the rats in New York City were resistant to the rat poison the city was using, most of the poison-resistant rats residing in East Harlem and the Lower East Side. A new rat-killing anticoagulant was soon developed, but recently, rats in England have become immune to this second generation of poisons. Jackson predicted that the same will happen in the U.S.: "Sooner or later, it's going to come." A few of the people representing companies that sell rodenticides got a little squirmy when Jackson said, "The use of poison is a failure of sanitation."

I had lunch with Stephen Franz, the vector specialist in the division of infectious diseases in the New York State department of public health, who worked on the Urban Rat Control Program in the 1980s. (We had pasta, as he's a vegetarian.) Franz has studied rats in India. He has designed huge city-size rodent-proofing plans; he advised on the rodent-proofing of the palace of Iraqi president Saddam Hussein when the U.S. was an ally of Iraq's. And he once managed a wild Norway rat colony of about six hundred rats in upstate New York, until the Urban Rat Control Program lost its funding and the colony was handed over to scientists who experimented with rat poisons. I listened to him describe high asthma rates in vermin-infested areas, and he talked about being in a building that was so full of cockroaches that you could hear them moving. "You're not supposed to be able to hear cockroaches," he said.

Also at the conference were representatives from Bell Labs, who were selling their latest rodenticides and offering samples of a new kind of snap trap. Some sort of barely hidden rodenticide manufacturing rivalry was going on—I got the feeling that some other rodenticide companies would not have come to the conference if they had known that Bell Labs, one of the conference sponsors, would be there with banners and key chains and pens and buckets of the rival rat poison. In the hotel fitness room, I ran on a treadmill alongside Al Smith, the national sales manager of LiphaTech, an American division of a giant French pharmaceutical company, whose U.S. headquarters are in Wisconsin, and another time later that spring, I even took him up on an invitation to visit the rodenticide factory, where I sat down with LiphaTech's plastic bait station, which is the Ferrari of rodenticide bait stations, with all of its rat-appealing corridors and the easy bait application features. I also saw what Smith called "the active ingredient." The active ingredient is what kills the rodents. In this case it is mixed with grains and molded according to various specifications and dyed a kind of aquamarine color that is meant to dissuade humans from eating it (rats don't see color). The active ingredient is kept in fifty-five-gallon-drum-size containers, enough to kill millions and millions of rats, in addition to other things. The active ingredient is mixed into the grain-based bait in a giant machine that looked like something out of an automobile plant. The men operating the machine treated the active ingredient carefully as they walked along the little aquamarine-colored trails of unpoisoned grain. I found the active ingredient to be a little scary; just standing near it made me nervous. On the way out, Al Smith mentioned some of the many clients around the country that used LiphaTech's poison baits, including a noted animal rights group.

But back in Chicago, I attended presentation after presentation, with titles like "Trapping Strategies for Rats and Mice" and "Poison Baiting Strategies." I sat among the pest control operators as they nodded and asked questions about droppings and mating and state regulations of rodenticides and the use of black lights to detect rodent urine. At the breaks I was able to hobnob with exterminators from all over the country, and in so doing, I started to get a feel for the rats of America. I met a man who got rid of rats at the Washington Monument, for instance; he talked about the history of rat infestations in the White House, and the historical controversy over the extent of rats in the White House (a revisionist school has tried to argue lately that the Nixon rats weren't that bad). I talked to a guy from the Pacific Northwest who kept his rat poison in plastic sandwich bags within the bait stations so that rats could still get it but slugs could not, an idea that appealed to another guy from the Southwest who had a similar problem with fire ants. I met a man who had killed rats at Mark Twain's house in Hannibal, Missouri—Les Shinn from Reliable Termite & Pest Control. The exterminator from Hawaii was dressed as if he were from Hawaii and told me that black rats love coconuts. I met a man who killed a lot of rats in a nice hotel in downtown Houston, and I met Bill Martinez of ABC Pest & Lawn in Austin. "I don't know what it's like for these places up North, but you get a dead rat in a wall in Austin in the summer and *ooooh-weee!* It *stinks!*"

After the hobnobbing, another presentation commenced, and a representative of a large pest control firm said, "The bad news is rodents are going to win this war against us humans. The good news is there's a lot of business."

A nice thing that happened to me was running into George Ladd, from Bonzai de Bug in New York. It was great to see a familiar face, and he recognized me right away. "Oh, man, it's great to see you, guy!" he said. "Have you talked to Bobby yet?"

ACTUALLY, I HAD ALREADY MET Bobby Corrigan, in a sense. In the interest of fairness, I should mention that I had talked to him once, briefly, just before the conference. I had called him and apparently caught him at a busy time, and he told me that rather than talk to him about rats, all I had to do was read his latest book, which would cover everything I needed to know. Of course, his book is entitled *Rodent Control,* and when I finally got my hands on a copy, I devoured it. I also read as many of his columns for *Pest Control Technology* as I could get my hands on, in addition to a story written about him when he was named one of the pest control industry's leaders a few years ago. So I already knew that he was born on Long Island, the son of an Irish laborer. "With eight brothers and sisters we were barely able to put food on the table, but when you are poor, you compensate by hanging out together as a family," he once told a *Pest Control Technology* reporter. "We realized that we had each other so we looked out for one another." He worked as a supermarket checkout clerk for two years before saving enough money to attend the State University of New York at Farmingdale, where he studied under another well-known rodent expert, Austin Frishman, whose lecture on structural pest control changed Bobby Corrigan's life—he switched his major from oceanography to pest control the next day.

Upon graduation Bobby worked for Fumex, a Long Island pest control firm, and had accounts all over New York City. After three years in the field, he went to Purdue University, in Indiana. Then, after a brief stint with Terminex, he returned to Purdue as vertebrate pest management specialist, a one-year position that turned into a sixteen-year job. He left Purdue to open his own pest control firm when his wife, a molecular entomologist in Purdue's entomology department, moved to Earlham College, a Quaker school, where she studied things like the DNA of finches. Today, they live on a seventy-acre farm in Indiana where they spend their spare time planting species of trees and grasses that are considered native and getting rid of the species that are considered invasive. Aside from studying bugs and animals on his farm, Corrigan enjoys writing poetry.

Reading Bobby Corrigan's book, one immediately gets a sense of why he is the superstar of the rodent control industry. First of all—and most obviously—he knows all about rats. He has studied them with a careful patience. And if he is not an expert on a particular area of rat infestation (rats in sewers, for example) then he is fully aware of the latest research (in the case of rats in sewers, Bruce Colvin's studies, which showed that rats prefer older, brick-lined sewers to newer ones, for nesting purposes). Just Bobby Corrigan's photos alone—of rat burrows, of greasy rat trails along walls and in ceilings, of rats peering nearly unseen from secret places—are enough to impress even the modest student of rats, and they would make a great wildlife calendar if people used photos of dead-grass-covered strips along a broken city sidewalk or of sewer holes or of splotches of rodent feces in calendars. Implicit in his work is the idea that there is no such thing as a monster rat. In *Rodent Control,* the rat is not evil. The rat is a rat. Of course, Bobby Corrigan understands that when an exterminator suddenly and unexpectedly finds himself face-to-face with a rat, it can be difficult to stay levelheaded. Bobby Corrigan has written, "Frightened you are—composed and clear thinking you are not."

But I believe that the secret to Corrigan's success is that he understands as much about the rat hunters as the rats; he relates to the man in the field, the guy with a can of roach poison on his back, who has been stuck in traffic all day and accidentally frightened the old woman in the upstairs apartment and is now looking down a toilet bowl as something rises up from the hole in the bottom. He writes of such a scene in the section entitled "A Rat in the Toilet Bowl," and he counsels the exterminator to stay calm, but fully understands that he or she may not be able to do so. Specifically, he recommends using a wild-animal loop snare, and placing the animal in a sack to bring it outside, but again, he confesses that he did not use a wild-animal loop snare the first time he saw a rat coming out of a toilet bowl. "As a novice pest control operator encountering a live toilet rat for the first time," he writes, "this author admits to first flushing the toilet and then eventually pinning down and crushing a toilet rat with the wand of a one-gallon compressed-air sprayer. It was neither a pretty scene or a 'professional event,' that's for sure."

Similarly, Bobby Corrigan understands the Sisyphean nature of extermination; while the persistence of rats is good for business in general, it can also be demoralizing, especially when the client has already paid you and expects you to keep coming back until every rat has disappeared. "Most pest management professionals and employees of warehouses and granaries often speak of their occasional encounter with 'smart rats,'" he writes in the chapter entitled "Challenging Rodent Situations." "And writing from experience, it once took this author three weeks of nearly daily effort to take out a single rat from a granary." Elsewhere he states, "In one government building in Washington, D.C., it took nearly one year before an elusive rat finally died of supposedly natural causes (old age). One of its legs was missing a foot, but the leg itself appeared healed over for some time. Perhaps this rat had lost its foot some months ago in a rat trap." Here Corrigan counsels that in certain special cases, where a single cagey rat just won't go away, a man with a rifle and night-vision gear may be the only way to eliminate it: "A sharpshooter quietly lying in wait at night is often used to take out the troublesome rat."

I spotted Bobby Corrigan just as the conference was about to begin, on the first morning. Approximately five feet and nine inches tall with light brown hair on the sides of his head and bald on top, with a mustache and wire-rimmed glasses, with a pen in his pocket and a name tag that said BOBBY, Corrigan was surrounded by rodent control operators who were lining up to speak with him, to shake his hand. His head twitched back and forth from exterminator to exterminator, each one's name instinctively on the tip of his tongue.

At the podium he had the exterminators immediately at ease, even laughing on several occasions. "Sometimes, people ask me, 'Well, how do I know I've got a breeding male?' Well, it's pretty easy to tell if you got a breeding male," he said. He was one with his audience. He didn't act as if he knew more than they did, even though he was the only person there who had photos of himself sitting amidst heaps of chicken dung in a poultry house watching and taking notes on rats as they passed by. His theme was the continuing education of a pest control technician. "When it comes to rodents, I hope all of us can agree, it's a constant learning experience." Everyone nodded his or her head.

"I've been out now for thirty years since I started a pest control route in New York City," he said. By now, some of the pest control operators had stopped taking notes and were just looking up at him in awe. "And I've worked in different environments with the mouse and the rats and whatever you have, and I realize how much we don't know and how much there's still to be discovered."

Bobby Corrigan spoke some more and monitored a panel discussion, but mostly he walked around the room surrounded by pest control technicians, answering hundreds of questions. And each time I approached him, a more aggressive questioner cut me off, so that at the end of his last talk I realized that, even though I'd come all the way from New York to meet Bobby Corrigan, I would go home without spending any one-on-one time with him, though obviously not without more knowledge on the rats of America. It felt unfair, in fact, to take any of Bobby's time away from the many pest control operators who were his longtime fans, his devoted followers. As I was pondering this, I looked over at the notes of the pest control operators sitting on either side of me—one from New Orleans, the other from St. Louis. They had walked away from the long desk to try to speak with Bobby, leaving their notebooks open, their last note-takings exposed. The one on my left said, "I fully agree with Bobby." The other one said, "Bobby was excellent!" The word *excellent* was underlined several times.

## Chapter 13

# TRAPPING

SOMETIMES, I CONFESS, as I sat in the alley late that summer and watched a rat emerge, as I studied its now predictable but still surprising path toward food, I felt an odd thrill of wild delight at the notion that I could perhaps myself catch that rat, trap it. It occurred to me that the rat *catcher,* spending his time in basements, dilapidated apartments, and alleys, is, in a strange way, a part of the rat's natural environment, more so than the average rat-avoiding citizen. Trapping would provide a means of observing a wild *Rattus norvegicus* rat up close. Not that I was anything of a trapper prior to my time in the alley, not that I had ever in my life hunted anything. In fact, I found the whole idea of rat trapping to be alternately exciting and horrific, appealing to both my lower and higher natures—though mostly to my lower, as my wife seemed to intuitively understand. Her first reaction was to utter, "Oh, God," followed, a short time later, by an admonition, "Just be careful," which I took as sagacious advice, and I also took as meaning that I should trap with a friend. Fishing is something that appeals to me; I have some experience in holding a rod, if less in bringing anything in. Thus, in ratting, I planned to utilize the catch-and-release method.

I'd spent some time on the telephone with a trap manufacturer discussing hypothetical rat trapping, trying to sound casual; I asked questions such as "If I was going to *maybe* have to trap rats, for example, what could I *maybe* use?" The saleswoman recommended chipmunk traps. Although I didn't say so, I felt that the rats in my alley would not fit in a chipmunk trap, much less get trapped in one. I couldn't decide if it was going to be easy to trap the rat and not necessitate buying a lot of stuff, or if it was even more difficult than I thought and I should buy a *lot* of stuff. In the catalog for Tomahawk animal traps, I found a lot of interesting trapping gear, such as Kevlar gloves, Kevlar sleeves, animal control poles, snake tongs and snake graspers, odor eliminators, Tomahawk Dura-Flex nets, telescoping syringe poles, dart guns, and even throw nets, all of which I could imagine being used in rat trapping. The goal would be twofold: (1) get the rat, and (2) stay the hell away from the rat. In the end, I bought a trap at my local hardware store. It was a toolbox-size rectangle with a tray in the middle intended to hold bait and a door on one end that dropped shut when the bait tray was disrupted. Manufactured by Havahart, it is designed for what are described as "nuisance animals." The motto on the box is "Caring Control for Small Mammals."

I FELT THE CONFIDENCE OF a cool night's air as I stepped outside with my trap, and after logging so many hours of observations, I felt as if I really knew the lay of the alley's land, as well as the alley's rats—as if trapping a rat was an actual possibility. I had arranged to meet with two friends, the artist and the poet who had been out ratting with me before. Matt, the poet, was going to pick up some bait at a deli on the way to the alley. I was meeting Dave, the artist, on the way there and he would carry my binoculars and notebooks and night-vision gear, while I carried the rat trap.

By 9 o'clock I was on the subway with Dave, carrying the large Havahart trap as well as the rope that I planned to rig to the cage to effect a release from a safe distance. On the way, we ran into an acquaintance, a guy whom we knew from high school who knew nothing of our rat endeavor—a central paradox of life in the city is that in the midst of several million people, each of whom seems to live life in complete anonymity, you can run into someone you know. As we greeted our friend, I moved the trap from my right hand to my left hand so that my right hand would be free to shake his. As I did this, I noticed that the guy looked down at the trap but didn't say anything. Then he looked up and said, "So what are you guys up to?"

I was practically bursting with my answer, of course. "We're going to try and trap rats," I said.

He looked me up and down and nodded hesitantly. Then he looked at Dave and said, "So, Dave, what are *you* up to?"

Dave and I came up out of the subway tunnel downtown, two blocks from the World Trade Center, which was still standing at that point. The weather was cool after a torrential rain, autumnally crisp. Near the alley, we met Matt, who had bought the bait as directed. In choosing the bait, I combined the expertise I had acquired while watching the rats eat garbage for a few months with a knowledge, culled from my rat readings, of foods that they have been reported to prefer—and then I just made some gut choices, so that in the end we ended up with sardines, Vienna sausages, and Kit Kat bars, all stuck together with peanut butter.

We were all excited to begin trapping. Yes, there were some unanswered questions. How would I release the rat after catching it? What would the rat do after I released it? Would the rat turn around and attack? Would we fall dead immediately from one of the many horrible diseases that rats can carry but do not always carry? And what was I thinking again, trying to catch a rat? Where was my self-respect, my instinct for self-preservation?

I was fortunate, however, in that Dave and Matt were a lot more relaxed about the whole operation. Though comforted by my relatively extensive rat knowledge, I am like the rat in the pack that is fine-tuned to smell fear. They encouraged me as I worked. "Put them *all* in there," Dave said, while Matt said, "That looks good enough." At last I had the trap set. I was ready.

THE ALLEY WAS ALIVE WITH rats that evening—rats streaking down the walls, from side to side, rats squabbling and screeching, rats eating trash. There were rats feeding on the Chinese-food side and rats feeding on the Irish-bar-food side. Rats were pulling themselves up through the holes in Edens Alley's cobblestones and racing around the corner and coming down the sidewalk. Watching the skitterings and the rat-happy spasms, I saw a nighttime symphony of movement, a chorus of disgustingness. From a rat's perspective, the alley was food-filled, maybe stress-free. I made a mental note to reestimate the population at some point—it seemed to have increased significantly.

With the baited trap in hand, I took a few steps into the alley. Immediately, I turned around and took a few steps back out. Then I collected my thoughts and tried walking into the alley once more, this time with Dave and Matt covering me—it felt good to have some backup, to have visitors in the alley. A few rats quickly scampered off as a result of our presence, but we were still and relatively quiet and many more rats stayed; several of the rats that were deep in the garbage bags continued foraging, as if we weren't even there. The rats who were en route from food to nest or vice versa returned to their nests for a few minutes but then resumed what appeared to be an abbreviated feeding pattern, which was roughly the same pattern only slightly more cautious. The rats seemed to be working around us.

I had some trouble rigging my safety release rope, the jury-rigged affair that I hoped would protect me when and if I caught one of the many rats scurrying around. I was kneeling on the ground in the rat alley, watching for rats around me in a paranoid fashion, and, as a result, spent a lot of time fumbling with my pocketknife. I was getting nervous about the whole thing—worrying about the flow of rats, worrying about jail, about death: rats can sometimes symbolize anxiety, for me, fear of the worst. Matt once again kindly suggested I relax. I looked up at Dave, who was smiling somewhat anxiously. I kept working, eventually choosing to place the trap at the highest part of the alley, a position that seemed to me at once out of view of the passersby on the streets and precisely in the course of the rats that were coming from Edens Alley down into the two trash areas to feed. I had seen a healthy amount of rat traffic along the wall, and in placing the trap in this path, I felt like a trained pest control technician, which, of course, I am not.

With my hands smelling like peanut buttery sardines (and with a savage urge to try one of the Vienna sausages), I left the alley and felt a surge of satisfaction. We all stood out in the street, near John DeLury Plaza, and looked into the alley with binoculars.

Ah, the excitement, the nail-biting and palpably semiwild thrill of ratting in the city!

The first rat to turn the corner from Edens Alley and head toward the trap elicited positive comments from all of us. We soon found ourselves talking to the rat, encouraging it, saying things like "Come on, rat" and "There you go, rat" and "Do you smell those sardines?" Unfortunately, the rat was extremely tentative; it took a step down the alley, seemed to notice the trap, then paused. As I have already stated, any pest control technician will tell you that rats are neophobic—i.e., fearful of anything new or different in their habitat. Matt and Dave and I watched excitedly as the rat made his decision and, after a few hopeful seconds, lurched and veered *around* the trap to head for the river of garbage, which, as a matter of fact, was growing, ever changing, ever the same.

It was spectacularly unclimactic as the next rat approached and repeated the maneuver, and in retrospect I feel ridiculous even thinking that a rat would choose my bait. This new rat made the same tentative investigation that had previously so tantalized us. And for rat after rat it was again the same: each rat seeming to note the trap, each rat perfectly avoiding the trap. They were proving precisely as wary, as sensitive to newness, as bait-shy, as reputed.

And yet there were so many rats in the alley and there was so much garbage—more bags had come out, more were coming—that I could not believe that not one would take our bait, despite knowing this hardened rat behavioral fact: when faced with a new food source, they will most likely stick with the old food source, until it runs out. Wasn't it possible that they would become acclimated to the presence of the trap in a few hours? Hoping for the best, I moved the trap to the other side of the alley, along the Chinese-restaurant-garbage side, in the midst of the Chinese food garbage. There was even more activity now. One rat seemed to climb up on top of the trap to investigate—or at least I think it did: it was difficult to see in the shadow of the long ridge of garbage bags. Notwithstanding, the result was the same.

In the hours around midnight we were serenaded by screeching cabs and sleepy-sounding garbage trucks, and at some point it occurred to me that rat trapping is not unlike fly-fishing—finding the perfect place in the streamlike alley and understanding the rats' garbage feeding preferences were both crucial. And as is the case for the fly fisherman when he stands in the cool, clear stream, our own sensory instincts heightened. We saw more clearly the flow of vermin and refuse, and I saw again—so much more so, in fact, that I was wondering how I hadn't noticed it before—that the alley was inclined, a hill. The source of this flow of rats, the stream of rodents, was a completely denuded peak, an alley-covered rise of the land.

It turns out that like trout, rats are incredibly skittish, wise-seeming, even, and hesitant to take a chance on the extraordinary sardine-juice-covered Vienna-sausage-and-Kit-Kat mix when a field of freshly discarded, partially eaten shrimp fried rice has already proven safe, and appears each night like the regular hatch of stone flies. After an hour or so, several juvenile rats were flirting with the trap, approaching it from the side, but then they too swam off to the plastic bags. The large, wary rats, meanwhile, seemed to grow even more suspicious: they paused, upwind, then stopped and started and finally raced past the trap and into the garbage. We were disappointed. Each time a rat checked out the trap, it still felt to us as if a rat could be trapped, *would* be trapped, just as each cast of the fisherman's fly rod brings new hope, fresh anticipation, until at last the fisherman becomes convinced that he is standing in the wrong spot or maybe fishing with the wrong fly.

So it was that at about one in the morning we collected the trap and broke up for the night. I felt I was close to understanding how to trap a rat, and yet I needed more studying and observing. As it turned out, though, I wouldn't be back down to see the rats in the alley for a while, because later that morning the World Trade Center would be destroyed—I can still remember looking up at the towers as we went down into the subway again. For the next few weeks, all of downtown was evacuated and blocked off.

*Chapter 14*

# PLAGUE

One reason that rats have a bad reputation is that they have been at the scene of some of humanity's greatest calamities, chiefly as carriers of the plague. Plague is often referred to as bubonic plague because of its symptoms, which include a fever and swelling of the infected person's lymph nodes, or buboes, followed by convulsions, vomiting, giddiness, severe pain, and dark spots on the skin. Death results from heart failure, internal hemorrhaging, or exhaustion. Other versions of plague are pneumonic plague, which is a kind of pneumonia, and septicemic plague, which is plague that invades the bloodstream so quickly that death can occur within twenty-four hours. The plague is also known as the Black Death, though it was not called the Black Death during the Middle Ages when it wiped out as much as 80 percent of the population of most towns and villages. It was first called the Black Death by Scandinavian chroniclers writing in the sixteenth century. Though the plague can cause parts of the body to turn black, when the Scandinavian writers used the term *black,* they used it to mean *terrible* or *dreadful* or *horrible.*

Plague can live indefinitely in communities of small rodents, such as marmots and gophers and various kinds of rats; rodents are considered the natural reservoir of plague. In fact, plague infects rats and kills them too, so that it could be argued that rats are as much victims of the plague as humans. When rats get the plague, they get it from fleas—most likely, a rat flea such as a *Xenopsylla cheopis*. A rat flea is about the size of this letter *o* and is shaped like a miniature elephant. A flea injects its trunklike proboscis into the rat to suck blood. When a rat flea sucks in rat blood infected with plague bacteria, the plague bacteria multiplies and eventually clogs the guts of the flea; the flea starves to death. In the meantime, before the flea dies, it feeds again and regurgitates the plague bacilli into the next rat as it feeds on the rat's blood. As many as one hundred thousand bacilli can be injected into a rat by a flea, but one plague bacillus could kill an animal as large as a monkey. When the rat dies, the flea senses the temperature change of its host and leaves the body of the cold, dead rat to find a warm, live rat. The flea then infects that rat, which either stays alive and breeds the plague for a time, or dies, causing more fleas to move on to more rats. Rat fleas prefer to feed on rats, and in areas where plague-infected rodents do not regularly come in contact with humans, there may be no human plague epidemics; the disease can live without consequence to man. But because rats live so closely to man, rat fleas will feed on humans (or any warm-blooded mammal) as a kind of second choice. The rat flea can wait a while for a human to appear; it can live for six months without a meal of blood. It can live in old rat nests or in fabric.

Plague epidemics begin when plague fleas begin jumping from rats to humans, as the rats die. Epidemics turn into pandemics when the disease spreads through wider areas, like a continent. There is a mention of what historians believe may be a plague epidemic in the Bible, thought to have occurred amongst the Philistines in 1320 B.C. The first plague pandemic swept the Roman Empire in the time of Emperor Justinian; between 25 and 50 percent of the population died. The Black Death of Europe was the second pandemic. It broke out in 1338, the end of a chain of events that may have begun with an infected community of a kind of large marmot, called a tarbagan, that lived on the arid plateau of central Asia in what is now Turkistan—a disease that would nearly wipe out whole cities originated in the most rural part of the world. It is theorized that the nomads who lived in the area were spared plague death because the fleas on the tarbagans were apparently repelled by the smell of the tribe's horses; a balance existed between flea-infected tarbagans and humans. Then something happened to the area that disturbed that balance. Historians speculate about earthquakes, but no one knows for certain. Another change occurring at the time was the building of a road, a great silk-trading route that connected Europe with China. Italians were especially interested in trading for silk, and they set up colonies along the eastern shore of the Black Sea, which was from where people like Marco Polo made their way to China. Along with silk and other trading goods, the traders brought back rats, probably black rats, which preceded *Rattus norvegicus* into Europe and were migrating along the human routes from Asia. First trading stations grew up along the routes, then towns. Unlike the nomads, the people in the towns became infected with the plague, probably due to infected rats.

The plague moved west along the silk route. It traveled with the human settlers and travelers and rats along the Volga River; it arrived on the coast of the Black Sea. David Herlihy, the plague scholar, wrote, "To spread widely and quickly, and to take on the proportions of a true pandemic, the plague must cross water. Contact with water ignites its latent power, like oil thrown upon fire." A famous plague story involves a khan of the Golden Horde, a Mongol state conquered and ruled by a grandson of Genghis Khan that got its name from the gleaming tent camp it set up along the Volga River. In 1347, in the Genoese Black Sea trading port of Kaffa, what is today the Ukrainian city of Feodosiya, the local khan battled the Italian merchants. The khan used a catapult to hurl plague victims into the Genoese port. The Genoese dumped the bodies but sailed back to Italy with plague-infected rats and fleas.

The plague quickly streamed west through Europe on and in the rats—the humans following their long-established trading paths, the rats following their long-established habit of following humans. It started in seaports. A Byzantine observer noted that the plague began in the ports and moved into the countryside: "A plague attacked all the seaports of the world and killed almost all of the people." People were dying as far west as Ireland, where, in Killkeney, a monk who described himself as "waiting among the dead for death to come" left blank pages at the end of his journal "in case anyone should still be alive in the future."

What was perhaps most black about the plague was that no one knew where it was coming from or what caused it or whom it would strike next. Science attempted to explain the plague but failed. When called upon, doctors at the University of Paris, the most exalted medical faculty in the medieval world, posited, citing Aristotle, that the cause of the plague was a conjunction of Saturn, Mars, and Jupiter, on March 20, 1345, at 1 P.M. "For Jupiter, being wet and hot, draws up evil vapors from the earth, and Mars, because it is immoderately hot and dry, then ignites the vapors, and as a result there were lightnings, sparks, noxious vapors, and fires throughout the air." Other indications that experts deemed plague-related were birds restless at night, frogs sitting huddled together, fruit becoming rotten and full of worms and falling from trees, the presence of unusual insects, large spiders, strangely colored gnats, ravens circling in pairs, mad dogs, and black vapors rising from the earth.

These observations made sense at the time because scientists generally understood infectious diseases as a matter of venomous atoms—invisible particles generated by rotting matter or emanating from people or animals or even objects already infected. Air infected by the venomous particles would become bad or *miasmatic*—i.e., poisonous. The venomous atoms were considered sticky and to be avoided. People avoided foul airs by holding flowers to their nose or dousing themselves with perfumes, which were invented as a response to plague. Some people felt that if they doused themselves with an odor more foul than the bad air, then they would be safe. Thus, in addition to bathing in rose water, people bathed in urine or stood for a long time in latrines. To some extent, such preventative measures often worked, though often not in the manner intended. In Italy, for example, doctors took to wearing robes made of *toile cirée,* a finely woven linen coated with wax and fragrance. Along with the linen robe, the doctor wore a hood and a mask and a long beaklike apparatus that was designed to filter the air. It made the wearer look like a large, sinister bird. In 1657, a friar, Father Antero Maria da San Bonaventura, who treated plague victims at a pesthouse in Genoa, noted that all the plague robe did was protect him from fleas, which the friar described as legion. "I have to change my clothes frequently if I do not want to be devoured by the fleas, armies of which nest in my gown, nor do I have force enough to resist them, and I need great strength of mind to keep still at the altar," he wrote.

A reader today might shake his head at the obliviousness of the flea-covered friar wondering why flea-covered people had the plague, but that reader has to put himself in the friar's shoes. It was not just what the friar noticed that mattered; how he thought about the world in which he noticed things mattered too. In *Fighting the Plague in Seventeenth-Century Italy,* Carlo Cipolla, an Italian historian who died in 2000, wrote, "We should not laugh . . . at the doctors of the scientific revolution." Commenting on Father Antero, Cipolla says that in the friar's mind the fleas were obnoxious but innocent; the friar's remark was meant as a casual attack on the linen robe and not on the scientific system that designed it or described the natural world. "Thus the system prevailed and the observation was lost," Cipolla added. "In the course of human experience thousands of brilliant and accurate observations must have gone astray simply because the related pieces of the mosaic weren't there. Thousands of other observations suffered no less sad a destiny. Accurate observations may be manipulated to fit into a faulty conceptual system with the perverse result of lending support to it. The examples that one could quote are innumerable and would cover the past as well as the present, the sciences as well as the humanities, religion and philosophy as well as politics."

Given the absence of precise knowledge about the plague, communities worried about contracting the disease used fear and malice to guide them. People blamed indecent clothing, corrupt clergy, and disobedient children. Some people blamed it on the general morality of the contemporary populace, including the fourteenth-century poet who wrote these lines:

See how England mourns, drenched in tears.
The people stained by sin, quake with grief.
Plague is killing men and beasts.
Why? Because vices rule unchallenged here.

People also blamed outsiders for the plague—visitors, immigrants, drunks, beggars, Gypsies, cripples, lepers, and Jews. Jews were accused of poisoning wells and springs, which in turned caused plague, even though Jews themselves were also dying of plague. In 1349, a Franciscan friar wrote, "And many Jews confessed as much under torture: that they had bred spiders and toads in pots and pans, and had obtained poison from overseas; and that not every Jew knew about this wickedness, only the more powerful ones, so that it would not be betrayed." Christians burned Jews all over Europe to prevent the plague. Jews were also confined to their houses and starved to death. Sometimes, they were spared if they converted to Christianity. Christians also blamed Islamic nations for causing the plague. Islamic people in turn blamed Christians.

The worst of the Black Death happened between the years 1347 and 1350, after which the second plague pandemic subsided a few times. It went from country to country and ended for good after it killed seventy-five thousand people in London. There are many different theories as to why plague ended in Europe: the climate may have gotten too cold for the oriental rat flea, or it may have been that the black rat, which lived closely with man in wooden structures in the cities, was replaced by the Norway rat, which, while also able to carry the plague, was at the time more likely to live in burrows on farms or at least a little farther from man. (The widespread development of sewers, today a natural habitat for *Rattus norvegicus,* was a century away.) Another reason cited for the end of the plague is that there may have been fewer fleas due to the increased use of soap. One recent theory suggests plague was not a rat-and-flea-related plague at all but an outbreak of anthrax, a disease that normally afflicts cattle but can cause symptoms similar to those described during the Black

Death.* Supporters of this theory say that there are no mentions of rats dying. On the other hand, there is no mention of rats *not* dying, and after spending a goodly amount of time with rats and exterminators, I can say definitively that people are oblivious to the true extent of rats, alive or dead.

Daniel Defoe's book *A Journal of the Plague Year* describes the last large epidemic of the Black Death as it attacked London in 1665. *A Journal of the Plague Year* is a novel but is also read, correctly, as an accurate account of the plague; Defoe, who was a small child when the plague came to London, immersed himself in contemporary accounts. In the book, Defoe describes the plague causing wealthy citizens to leave the city for their country homes and the plague ravaging the low-income neighborhoods: the great plague of 1665 was known at the time as the Poor's Plague. He describes camps of people held at the edge of the city, distrusted foreigners. He describes fear as it infests the city; he details people plundering and looting abandoned homes, people taking advantage of other people's dire circumstances. There are numerous examples of cures designed solely with profit in mind. "Incomparable drink against the plague, never found out before," one handbill said. Defoe writes, "[T]hey not only spent their money but even poisoned themselves beforehand for fear of the poison of the infection . . ." Even people with honest

* David Herlihy supported this theory, as did Dave Davis, to some extent. Davis researched the record of rat mentions during the Black Death. But I have talked to plague experts who now dispute the anthrax theory, and when I read the ancient notetakers, I tend to believe the disputers. A biologist who investigates plague for the federal government told me that he had investigated ancient Roman crop records and seen that when the crop returns increased, a subsequent increase in plague cases followed. Also, just because medieval accounts don't mention rats per se doesn't mean they aren't referring to rats or ratlike creatures. Here is an example of a medieval naturalist, who pointed out that plague came "when snakes, bats, badgers, and other animals, which dwell in deep holes in the earth, come out in fields in great multitudes and forsake their ordinary dwellings." I count Norway rats as among possible hole-dwelling creatures. As for black rats, which would have lived in dwellings in the medieval cities, I think that people might have had them and not mentioned them. As any exterminator will tell you, and as I mention above, people would not believe the number of rats that are around them every day.

intentions unintentionally exacerbated the situation; in an attempt to stop the contagion, cats and dogs and even rats were killed. Thus, fleas would more rapidly have jumped to humans.

Defoe describes the city of London reacting like an organism itself, an organism that, while not devoid of good and selfless impulses, is also governed by hunger and fear. At the end of the plague's year, London manages to hold itself together. The city survives the danger brought upon it, but as a hero, the city is flawed, simultaneously noble and base in the face of mysterious danger. "The conclusion reminds us of the moral of Thornton Wilder's *The Skin of Our Teeth*: man comes through his ordeals and tests but only just," Anthony Burgess wrote in an analysis of Defoe's book. "In *Robinson Crusoe* man builds a community from scratch. In the *Journal* Defoe asks whether man can do more than build: can he preserve as well? We are doubtful when we see how badly some of the citizens behave towards each other, but, when we have added all up, we must conclude that the city has done rather better than we expected: it has gained no high marks but it has certainly passed. This is in conformity with Defoe's qualified liberalism, which means a kind of optimism. It is neither God's grace nor innate goodness which saves man's soul alive; it is rather his need for the community, his concept of the desirable life as one lived collectively."

New York City felt more like an organism than ever to me that September, and if it is a living, breathing, horn-honking, and fume-emitting organism during noncrisis times—pulsing with human traffic that wakes up at early-morning rush hour and slows down when the cell-like humans mostly go to sleep—then the city was an organism that was momentarily on life support after the World Trade Center was attacked: the National Guard stopping the flow of panic-ridden pedestrians, the streets blocked off by more-heavily-armed-than-usual police, sewer lines and gas lines and water lines cut and leaking and destroyed—its vital signs suspended or on hold. At the site of the World Trade Center itself, vast baseball-stadium-esque lights made the night seem like day, and cranes worked over the giant, smoke-filled hole as if they were doctors standing over an operating table, over the city's huge and still-bleeding open wound.

Reading Defoe while living in New York after the World Trade Center was destroyed made me realize that a time of crisis can sometimes offer up a test of a city's abilities—for while you inevitably see some of the worst things that people can inflict on other people, you also see some of the best. During those darkest days there were stories of people taking advantage of people, of looting and theft, but there were also moments of tremendous human generosity, examples of the resilience of the city, of people feeling the happiness of being with other people. When the twin towers collapsed, a small Greek Orthodox church across from the World Trade Center was destroyed, and I remember that the archbishop of the Greek Orthodox Church stated afterward, "We have seen the abyss, the ugliness and darkness of evil. In what followed we have seen the beauty and brilliance of good."

*Chapter 15*

# WINTER

I DID NOT GET to my rat alley for a long time that fall; streets were blocked off all over downtown. A sense of dread hung over the city for a long time, though eventually it lifted, not coincidentally, I would argue, with the ability of the populace to walk through the streets again, to go to work and buy lunch, pick up a newspaper or have a drink. Naturally, I was interested in seeing my alley again, and after my own city life regained some routine, I even began thinking, given my own strange area of expertise, of rats. Fear was what the city was facing—with threats, real or perceived, on its buildings, its bridges, its infrastructure, with concerns for family and safety and life—and rats are like fear: they creep in at night, in those unseen places of vulnerability. There was a time in New York, in the 1920s, when scientists proposed a great wall along the waterfront to shut out rats completely, to seal out rats and, thus, forever end rat fear. Eventually, though, the idea was deemed implausible and abandoned: rats will always get through. In those days when I was unable to get back to my rat alley because of blockades, it dawned on me that rats would get through to the streets that were cut off from people and maybe thrive. It dawned on me that there might be a lot of rats, actually.

Of course, I didn't mention this to people. Nobody really wanted to hear about rats, and who can blame them? The newspapers and television news programs weren't concerned with rats, to be sure. Rats were even censored to some extent. A movie entitled *The Rats*, originally scheduled to premiere on television the week after the disaster, was canceled; it was about New York City being overrum by rats, packs of them attacking residents.* As it happened, I was actually glad to hear that pest control officials everywhere were thinking about rats—it made me feel more sane. Eventually, I began hearing from exterminators that I know, and like small-business owners all over the city, some were adversely affected. George Ladd of Bonzai de Bug had a large construction job that he'd been counting on canceled after the disaster. On the other hand, Barry Beck was getting more work. When I talked to him, he was on his cell phone a few blocks from the site of the building and said he thought the rat population was increasing dramatically downtown: "They've proliferated. They've compounded, multiplied, and intensified," Barry said.

Late that fall and into the winter, Barry was working overtime all around the World Trade Center, at the small businesses and restaurants that surrounded it. He said he was forced to keep a full-time rodent control crew at the office of one large business located near the World Trade Center site. I found a strange comfort in hearing that Barry Beck was still on the job. Newspapers reported on whether large

---

* Instead of showing *The Rats*, the television network showed an old movie, *The Nutty Professor*. One of the few places that did mention rats was the *New York Post*, though, in my opinion, it tended to overmention them; like something out of Defoe, it equated rats at every opportunity with foreigners and to American citizens who consorted with foreigners—the paper used rats, in other words, to sell rage and xenophobia, which in turn sold papers. Some examples of the numerous *Post* headlines using the word *rat* are THE RAT'S BACK, RATS HELP TRAP OTHER RATS, MESSAGE FROM THE RAT HOLE, RATS JUMP FROM SINKING SHIP, OUR SPIES CLOSE IN ON EVIL RAT HOLES, and RATS GALORE. A political cartoon featured a giant rat that was labeled "terrorism" and about to eat a giant piece of cheese on an American flag that doubled as a giant rat trap. After an American was captured in Afghanistan when American military forces invaded in retaliation for the destruction of the World Trade Center, the front page of the *Post* was like an overgrown rat haiku. It read:

LOOKS LIKE A RAT
TALKS LIKE A RAT
SMELLS LIKE A RAT
HIDES LIKE A RAT
IT IS A RAT

When I went to the New York Public Library to look at this issue of the *Post* for a second time, the librarian who handed it to me said it was repeatedly requested. When we finished speaking, a German tourist politely inquired as to whether the library's archival copy was for sale.

multinational conglomerates and financial banking firms would stay or go, but for my own part I felt it said something good about the city that Barry Beck was sticking around to fight rats.

There was some controversy over precisely where the rats were. Some people thought the rats were going into what remained of the old World Trade Center site itself. Barry disagreed. "My personal feeling is that they've left the World Trade Center area and went into the surrounding buildings," Barry said to me at some point. "That's my personal feeling. I think they went into the restaurants and things."

When I did finally go out at night again, I put off going to Edens Alley; I looked for rats elsewhere first. As I walked, I noticed right away that someone had put rat poison all around; there were bait stations filled with rat poison everywhere. I saw a number of them on Thames Street, for instance, a short, alleylike street that runs off lower Broadway and connects to the World Trade Center site. Rats were streaking up and down Thames Street, stopping in the bait stations momentarily to feed on poison. (Thames Street was where trucks unloaded food donated to the World Trade Center rescue workers by people around the country—lobsters came from Maine, for example, and Cajun food from New Orleans, the scraps of which no doubt helped feed the rats.)

And then when I checked in on Theatre Alley—my second favorite rat alley, the alley that was the site of the famous rat attack and is hidden right across from City Hall—it was as if I was checking in on an old friend. How happy I was to see that the row of old buildings in front of it, as dust- and ash-covered as they were, had survived! The alley itself felt protected, safe, secretly secure from the apocalyptic event two blocks away. Inside the alley itself, where I had once seen scores of rats, now there were only bait stations, lined up and down the walls. I checked the poison identification labels and discovered they had been laid out by the city's rodent control department. Then, as I continued to look around, in the streets immediately surrounding the World Trade Center and even as I looked over the remaining police barricades, I saw bait stations in other alleys and even lining whole streets. It was a tremendous show of antirodent strength.

I GOT IN TOUCH WITH some people I know at the health department's rodent control office. I talked to Dan Markowski. Markowski is a tall, ponytailed Southerner, who works in the section of the health department called vector control. A vector is anything that carries a virus or a disease. Rats are excellent vectors: rats are vectors of plague, as well as diseases such as typhus, salmonella, rabies, hantavirus, and leptospirosis, which is sometimes called yellow jaundice or Weil's disease and is spread by the rats' infected urine, usually in water. (A few years ago, a dog in Brooklyn's Prospect Park died of the disease after he ate a rat he caught in a big puddle.) Many insects are vectors, and New York's health department monitors the population of New York ticks, which can transmit Lyme disease and Rocky Mountain spotted fever, and mosquitoes, which can carry typhus and malaria and West Nile virus. Roaches are also suspected of being a source for asthma, as are rodents.

Dan Markowski had studied ticks in Rhode Island, where he went to graduate school. After graduation, he worked with mosquitoes in New Jersey, on the Jersey shore. When he came to New York, in 2001, he was mainly responsible for mosquito work; he spends a good deal of time testing mosquitoes and considering spraying insecticides from helicopters. But he was also interested in a project that involved live rat trapping, something New York's rodent control office had abandoned decades before but was considering starting up again. "It's pretty exciting," he said.

Before trapping could begin, some basic, if extensive, sanitation work had to be done downtown. After September 11, Markowski inspected downtown restaurants that had been closed due to the quarantine and left unattended for many days. The condition of the restaurants was especially bad in the blocks surrounding the World Trade Center. At one restaurant, on Murray Street, Markowski and his colleagues had to put on protective suits and breathing apparatus to go in. The restaurant had had a long self-serve buffet station out on the morning of September 11 when everyone had evacuated. The buffet station had been left alone for weeks. A fireman who had been working in the WTC noticed Markowski and the other health department officers as they went into the restaurant. "I don't want your job," he shouted over to them.

The stench that hit them as they entered the restaurant was disgusting. "It was just this overwhelming, horrible odor that was just as bad as anything you could ever imagine," Markowski remembered.

They moved farther in. One of Dan's colleagues turned to him. She was also wearing a suit. "Dan," she said, "I've got to get out of here. I'm going to throw up."

Next, Dan thought he was going to throw up too, but he didn't.

When they made it to the buffet counter, they saw that all the items offered had putrefied. "There was just this slurry or sludge of rancid stuff," Dan said.

When some of the bags broke as they dragged them out of the restaurant, the emergency officials outside the restaurant scattered. Sanitation workers carted them away.

IT WAS DURING THESE PUTRID food removal forays that the health department began to think about rats. A fireman who noticed their health department jackets pointed to the partially destroyed building at 5 World Trade Center. "You guys better get in there," the fireman said. Dan and his boss, James Gibson, the director of the vector control department, realized that rats might become a big problem.

They began inspections. Strangely, there were not a lot of signs of rats on the streets nearby. Health officials suggested that the rats may have been finding food underground, in the basements of the buildings, in the underground restaurants. "You had all of these restaurants with all of this food, so there was the food source right there," Dan said. He paused as he continued, uncomfortable with what he was about to say, because another possible food source was the bodies of the people killed in the collapse of the towers—as public health officials, they had to think about such things. "And then you had a lot of deceased individuals," Dan said. "The thing is, that could be a food source."

When Dan said this, he paused and spent a few seconds trying to look like a professional health official.

Among the people who volunteered to help at the World Trade Center were the employees of Terminix, one of the largest pest control firms in the country. They were certainly not as celebrated as other kinds of workers who volunteered at Ground Zero, but they were just as happy to help. (LiphaTech, the company I visited in Wisconsin, donated bait stations.) "We had more employees volunteer than we had places to use them," said Mike Baessler, a Terminix executive. Baessler flew into the city from Memphis.

"We didn't do rodent-proofing," Baessler said. "Around Ground Zero that was impossible."

They put out a thousand bait stations, just in the World Trade Center, an enormous amount. They put 115 steel-wire bait holders in the area's sewers. They worked long shifts, from dawn to midnight, trying to stop rats.

"We thought, let's create this perimeter around Ground Zero," Baessler said.

Finally, with the Terminix volunteers, the health department arranged to be escorted into the basements of the World Trade Center, the underground rooms of the area—that is, the ones that had survived. Again, they did not see live rats. They did, however, see a lot of evidence that rats had been there. The rats appeared to be avoiding the rancid food, as they had predicted, though they were obviously eating what they could—most of a cookie store, for instance. But in the tunnels, in the inch-thick dust, in the lights of their flashlights, the rat fighters couldn't believe what they were seeing: there were thousands and thousands of rat tracks.

"The dust was just totally run through with rodents," Baessler said.

"The rodent population may have been tremendous," Dan recalled.

If they ever worried that they had put out too much poison, they remembered those tracks.

They circled the old building site with bait stations; they placed rodenticide all up and down the nearby streets. People in TriBeCa and in Battery Park complained of the rats in the streets that winter, and there *were* rats—I saw them. But there would have been an awful lot more rats if it hadn't been for the health department. Months later, James Gibson testified to the city council that they had successfully contained a potential rat explosion—a statement that was difficult to prove, or even believe if you didn't fully understand the downtown rat situation.

Dan Markowski summed up the exterminators' dilemma this way: "If we're out there doing our job, there are no rats. But it's notoriously hard to confirm our job. The only way to prove it is to stop." I think of them as being like spies, or undercover police, the people charged with securing the invisible perimeters who work in the areas where no one else goes. When they are doing what they are supposed to be doing, you don't always know. You only hear about them when something goes wrong.

After so many weeks away, I finally got a chance to spend some time back in my alley. It was the beginning of the winter, and it was a cold, rainy night. I came across the Brooklyn Bridge and looked down on the nighttime water, the dark waves; at the crest of the bridge, I could see the lights at the sites of the destroyed towers, glowing all throughout downtown like a steady fire. I passed City Hall and City Hall Park and heard the starlings that always sit in the London plane trees; they chattered busily as if nothing had happened. I walked toward Theatre Alley, the buildings completely protecting it still all covered with soot and ash. Aside from being filled with rat poison in the health department's secret rat offensive, the alley had been sanitized: the vacant lot had been covered over, the rubble-filled rat hole sealed, and the trash in the alley had been cleared out. The rat's natural habitat had been destroyed. I stood there for a minute wondering what could have happened to Derrick, the guy who called himself Rat Man, who had nearly had all the rats trained. There was a lot of talk about the difficulties faced by people who lived downtown after the World Trade Center was destroyed, but I never heard mention of the fate of the people who lived in the street downtown. Later, my friend Matt saw him recycling cans in Greenwich Village.

It was still raining by the time I got to Edens Alley, and I was feeling down about everything. I didn't have much hope for the rats. I also had mixed feelings about having any hope for rats.

I took the long way and walked past the Fulton Fish Market. The market was closed up—but all of a sudden I though I heard a flute. It was coming from inside the market. I followed the sound, peering in each of the fishmongers' stalls. The floors were clean; the place had been hosed down. But the place smelled like fish. I saw a man sitting alone on a forklift playing the flute—a piper. He saw me and stopped. "Keep playing," I shouted.

"It really carries," he said. He started up again.

I listened for a while and then went around the corner to my alley. I was afraid to look in, and when I did, I didn't see anything. Dejected, I looked in again, and then I saw the gray streak, the blur. It was a rat, followed by rats. I didn't see any poison. The rats in my alley were still there. And, as I would learn, they were pretty much forgotten.

*Chapter 16*

# PLAGUE IN AMERICA

FEAR, WHICH THE RAT, more than most creatures, so impressively inspires, is a wild thing, and it can turn a man into an animal, direct him toward his basest impulses, his lowest nature. I mention this because as New York was considering its situation—repairing itself, rebuilding, reorganizing some of its civic functions and rehabilitating itself on the whole—I was still thinking about rats and plague. Plague came to America at a time when it could have been prevented from spreading but wasn't, because of fear. The very first plague case in the continental United States appeared on March 6, 1900, in San Francisco's Chinatown—by the Chinese calendar, 1900 was the year of the rat—and it was the same plague that had killed senators in ancient Rome, that had killed kings in medieval Europe. This time, though, scientists not only understood that it was transmitted via rats but had even discovered methods to combat the spread of the disease. Fear kept them from utilizing that knowledge—fear on the part of the city's business interests, fear that in turn inspired fear in the poorest parts of the city, which were most susceptible to disease and its ramifications.

The plague that arrived in San Francisco was part of the third plague pandemic that had broken out in China in 1850. Alexandre Yersin, a French microbiologist, identified the plague bacillus that was eventually named for him, *Yersinia pestis,* in 1894. Yersin worked with Louis Pasteur at Pasteur's institute in Paris. Yersin had met Pasteur after Yersin had cut his finger while operating on a man who had been bitten by a wild dog; his finger still bleeding, Yersin ran immediately to Pasteur's laboratory, where he was vaccinated with Pasteur's new rabies vaccine. When a plague epidemic erupted in Hong Kong, Pasteur sent Yersin to investigate. Yersin wanted to draw fluid from the enlarged nodes of the plague victims, but he was not allowed access to the morgue. On the advice of an English priest, Yersin bribed two English sailors working at the morgue for access. He drew fluid, looked at the microbes under a microscope, and in his journal wrote of the discovery that had eluded scientists for centuries: "This is without question the microbe of the plague."

After Yersin discovered the plague germ, he looked at the world in a whole other way; suddenly, he noticed the dead rats—all around the hospital and Hong Kong. He discovered that these rats were infected with the plague. Investigating further, he discovered that the people in China's mountain villages had long known that the plague outbreaks were preceded by rats dying. At this point, Yersin still did not suspect that fleas transmitted the plague from rats to humans. That link was made by Paul-Louis Simond, another scientist from the Pasteur Institute, who went to Vietnam to treat people with an antiplague serum during an outbreak there. He saw further evidence of rats' connection to the disease when he learned that workers at a wool factory had come down with the plague after being forced to clean up dead rats. "We have to assume," Simond wrote, "that there must be an intermediary between a dead rat and a human." At another plague outbreak, Simond began experimenting with rats in cages in his tent. In a jar, covered with a fine, flea-proof mesh, he hung a healthy rat in small cage just over a rat dying of plague. When the plague rat died, the fleas jumped to the healthy rat, which died a few days later. As a control, Simond placed a flea-free rat dying of plague in a jar with a healthy rat. The healthy rat stayed healthy.

When Chick Gin was found dead in the basement of a San Francisco flophouse, on that Monday morning in March 1900, an assistant city physician noticed swollen lymph nodes in the decomposed body. Lymph was extracted and taken to a federal quarantine station on Angel Island in San Francisco Harbor, where it was inspected the next morning by Dr. Joseph J. Kinyoun. Kinyoun had come to San Francisco after setting up the first infectious-disease laboratory in America, on Staten Island in New York City. He was a physician with the Marine Hospital Service, a predecessor to the U.S. Public Health Service. He was a young hotshot, maybe a little full of himself, though he had some reason to be: he was one of the few men in America to have been to Paris, Berlin, and Vienna to study infectious diseases. He had worked with Pasteur and with Kitasato, a Japanese microbiologist who had simultaneously discovered the plague bacillus independent of Yersin. Kinyoun was the right person in the right place at the right time, though he considered it a demotion to be moved to San Francisco, where he lived with his family on Angel Island, which was to him a bay-locked rock.

In San Francisco, on the day after Chick Gin's body was discovered, Kinyoun examined the bacillus and injected it into three animals: a guinea pig, a monkey, and a rat. The microbe looked like plague. Kinyoun reported his concerns to San Francisco's board of health. At first the board cooperated, enacting Kinyoun's call for a quarantine of Chinatown. Kinyoun's success in winning a quarantine did not inspire the Chinese residents of Chinatown, who referred to him as "the wolf doctor." Chinese residents feared for their lives and property. Police were sent in to keep the quarantine. When plague had arrived in Honolulu a few months before, the officials there were so intent on saving the city that they considered burning it down and ended up burning much of Chinatown. (The outbreak of plague in Hawaii is sometimes called the second-worst disaster in the history of the state, after the bombing of Pearl Harbor.) Non-Chinese residents of San Francisco were already calling for the destruction of Chinatown, an idea that racist politicians were happy to support; they used fear as an accelerator for their cause, which was hate. The *Call* said, "Clear the foul spot from San Francisco and give the debris to the flames."

The business community was terrified that plague would translate into a boycott of San Francisco goods, that tourists would stay away from San Francisco, that railroad business would suffer. The next day, under pressure from business leaders, the newspapers called the plague discovery a scare. They wrote that it was a plan executed by a corrupt board of health to make money. PLAGUE FAKE IS PART OF PLOT TO PLUNDER, said the San Francisco *Call*. The papers accused the board of health of seeking payoffs in order to, as the *Call* said, "get snout and forelegs in the public trough." Similar stories were published in the *Chronicle* and the *Bulletin*. The only daily paper that acknowledged the real possibility of plague in the city was the *Examiner*. The *Examiner* was owned by William Randolph Hearst, and Hearst decided there were more newspapers to be sold by playing up a plague scare than playing it down. A building trades journal called *Organized Labor* also spoke up for an investigation of plague, but it was merely using plague as an excuse to vent more anti-Chinese hostility. "Brothers, wake up!" *Organized Labor* said. "This is a matter of vital importance and should receive thorough consideration in your meetings. The almond-eyed Mongolian is watching for his opportunity, waiting to assassinate you." Many reports tried to make plague sound like an innocuous disease, like mumps. A poem in the *Bulletin* ended like this:

> And the advertised, boasted bacillus
> Is a gentle domestic concern,
> And the doctors who fill us and pill us
> Have libeled it sadly, we learn

Eventually, under pressure from businesses, the government lifted the quarantine.

Then, the guinea pig, the monkey, and the rat died.

ONE OF THE THINGS THAT the rats that were spreading plague did to San Francisco—aside from bringing the plague to America—was to cause one population of the city to see the way another population in the same city was living. Shortly after the guinea pig, monkey, and rat died, Mayor James Phelan reluctantly organized one hundred volunteer physicians to search for plague victims in Chinatown, a twelve-block area where twenty-five thousand Chinese people lived. When the physicians went in, they were explorers in another land. The doctors were shocked at the conditions they discovered. There were mazes of holes and secret tunnels connecting homes. In the underground rooms, holes were cut in sewer pipes in lieu of bathrooms; when the sewer pipes filled up, sewage backed up in the underground rooms, under row after row of bunk beds. And there were rats. The inspectors themselves complained of odors that made them nauseous as they did their work. They could not find any plague victims, however. Chinese residents, concerned that their homes would be burned down, hid their sick relatives and then shuttled them out of the city in small boats at night. Sometimes, when an inspector arrived before a body could be removed, a dead man would be propped up next to a table in an underground room, his hands arranged carefully over dominoes.

The newspapers continued to deny the presence of plague; they emphasized disputes over Kinyoun's diagnoses. People did not want to believe Kinyoun, on the one hand, and on the other, the bacteriological approach to medicine was still new. Kinyoun had worked with the latest in scientific equipment at the Pasteur Institute, but in San Francisco doctors considered swollen lymph nodes to be a sign of venereal disease and did not necessarily use microscopes. In fact, most physicians in San Francisco still saw human infection as a result of the inhalation of bad airs—a belief left over from the Middle Ages. The U.S. assistant secretary of agriculture at the time reported that Asians were particularly susceptible to plague because they ate rice and were deficient in animal proteins.

The *Examiner,* meanwhile, was practically exuberant about the plague, the journalistic equivalent of someone yelling fire in a movie theater. The *Examiner* published panic letters, and an editorial described the "invasion." In New York, the *Journal,* also owned by Hearst, published a special edition titled "Plague Edition," which was delivered to cities throughout the country. The headline was BLACK PLAGUE CREEPS INTO AMERICA. The accompanying, commissioned painting depicted men collecting bodies in the street and people dying from fright. The *Journal* predicted "plague-stricken men and women, out of their minds with pain, rushing naked about the streets." Because they did not have details from the plague epidemic yet to be officially declared in San Francisco, the *Journal* writers cribbed from Daniel Defoe's *Journal of the Plague Year.*

The *Examiner* only stopped when the other newspapers began attacking it. The *Bulletin* called for the inoculation of the *Examiner* with plague bacilli. "It should be removed," the *Bulletin* wrote, "this city would be healthier, corporeally, morally and politically." Daniel Meyer, a well-known financier, attacked Hearst directly. "It is the nature of the man to tear down," he said in the *Bulletin*. The *Bulletin* reported that thirty thousand tourists had been driven away from San Francisco by the plague stories. There were reports of cargoes being sent to other ports. Now, Mayor Phelan worked to deny the existence of the plague; he sent out letters to cities around the country asserting that everything was fine in San Francisco. Newspaper publishers met and agreed not to mention a quarantine on Chinatown. On April 1, the *Examiner* stopped publishing plague news. Meanwhile, more people in Chinatown continued to die from plague.

Fearing a Black Death-like health crisis, Kinyoun wired Washington, D.C., to say that the city was facing an epidemic. General Walter Wyman, the surgeon general, ordered more health service officers to San Francisco. The surgeon general convinced President William McKinley to apply a quarantine on all people of Asian ancestry leaving the state of California, such that they could not leave without certification from Dr. Kinyoun's Marine Hospital officers. The navy patrolled the harbor in armed boats. The Chinese Six Companies, a group representing Chinese business interests in San Francisco, sued in federal court and had the quarantine lifted, arguing that as designed it applied only to interstate traffic and not to travel within California and that it denied equal legal protection to the Chinese community.

Frustrated, Kinyoun asked the Chinese to submit to inoculation with an experimental preventive drug. The Chinese Six Companies agreed. But then non-Chinese doctors in Chinatown began spreading rumors that the drug had killed people. A crowd gathered before the offices of the Chinese Six Companies, people crying out for the company's officers to be inoculated before anyone else was; the crowd wanted the business leaders to be their lab rats. The business leaders refused. The crowds were on the verge of a riot. People decided to resist the injection. When a team of Kinyoun's physicians went through town to inoculate, all the businesses and residences were closed.

It was a standoff of paranoias, a fear face-off. The board of health debated moving all Chinese residents to a detention camp on Mission Rock, a small island in San Francisco's harbor. Word went out in Chinatown that anyone seen going to such a camp would be killed: checking into a plague detention camp would be tantamount to admitting the existence of the plague, which many Chinese residents wanted to deny. Meanwhile, doctors were being offered large sums of money by business interests to show that plague did not exist in Chinatown. The *Bulletin* ran a picture of the members of the board of health and suggested they be exterminated; a large headline read: THESE MEN ARE MARKED. Health officials fighting the plague were referred to as "the perpetrators of the greatest crime that has ever been committed against the city." California's governor, Henry Gage, worked hard to deny the plague. He assailed "plague fakers." He proposed life imprisonment for anyone claiming there was plague in San Francisco. He suggested that Joseph Kinyoun had planted plague bacilli on the Chinese man who died. Soon, all sides could agree on one thing: Dr. Kinyoun was a problem.

The attacks against Kinyoun were notably malicious and slanderous even in a town with a long history of yellow journalism. Kinyoun held fast; his arrogance made him immune to some extent. He turned down bribes. He went on trial in the city for contempt and was eventually found innocent. He was constantly being lampooned in cartoons such as the one that showed him being injected in the head with plague serum. His work was described by the press as "stupid and malignant." Meanwhile, he lived on the desolate island in the bay, with his wife, who was also unhappy there, and their children. He came down with an ulcer.

"Kinyoun is to go," the *Chronicle* editorialized.

In February 1901, after a new team of scientists arrived from Washington, D.C., and confirmed six more plague cases, Governor Gage met secretly in Sacramento with the heads of the railroads. They agreed that no newspaper would ever mention that the plague had ever existed, and they orchestrated an Associated Press dispatch announcing: "[T]here was not now nor has there ever been cases of bubonic plague in California." They also sent a group of newspaper publishers to Washington, D.C., where the publishers met with the president and California's senators. They all agreed to clandestine plague eradication measures and arranged to have Kinyoun fired. At the end of March, Kinyoun, America's preeminent infectious disease expert, was reassigned to Detroit, Michigan.

That spring, after the governor of Texas sent a telegram to the surgeon general threatening a quarantine of California if the plague was not contained, Chinatown was quietly fumigated with sulfur. (The health inspectors boasted that they had used a minimum amount of sulfur to do the job.) The Chinese Six, though they had previously admitted plague in Chinatown, now said that no cases had ever existed. Again, inspectors could find no bodies. After sixty days of inspections by state health officials, the governor shut down antiplague operations. The U.S. Surgeon General closed its federal plague eradication office. For eradicating the talk of the epidemic, Governor Gage was lauded as "the people's friend."

Then, on July 5, a Chinese undertaker accidentally brought the body of a plague victim to Dr. Rupert Blue. Blue had been a colleague of Joseph Kinyoun's and he'd replaced Kinyoun at Marine Hospital Service, and if Kinyoun was arrogant and imperious, then Blue was smooth and accommodating, a politician. Blue investigated. He pronounced the case plague. More plague cases were discovered—

on July 8, there were four deaths within forty-eight hours, in a single Japanese household. State officials called it sewer gas poisoning, then charged Blue with inoculating the victims with plague. It looked as if Blue might be reassigned to another city as well, but this time the Japanese community, as opposed to the Chinese Six, would not keep the plague secret. There was a new mayor, Eugene Schmitz, the former first violinist of the San Francisco symphony, and initially he was also ready to play along with business interests—he fired health officials who insisted there was plague, and he refused to print plague statistics—but now health departments and governors from two dozen states protested California's handling of the situation. More plague reports made their way East. Health officials from around the country met in Washington and passed a resolution against the "gross neglect" of the California Board of Health and the "obstructive influence" of Governor Gage. The states threatened a national quarantine against California. At last, plague eradication efforts began in earnest. Wooden floors were ripped up to reveal years of broken sewer pipes and cesspools. Rats were trapped all over San Francisco. The last case of plague was identified on February 19, 1904.

In 1906, after the Great San Francisco Earthquake, there was another plague outbreak in San Francisco, but the city had learned its lesson. The federal government immediately set about cleaning up the city and trapping rats. Rupert Blue was put in charge of plague eradication. This time none of the cases were in Chinatown, which may be one explanation for the quicker treatment of the disease. Officials noted that the city seemed to be infested with rats, and since the last outbreak, more and more scientific studies had confirmed the link between human plague cases and rats and rat fleas. Rat catchers used poisoned molasses and bread laced with arsenic to kill rats until two children died after ingesting the bait, at which point rat catchers used traps. To encourage the public to catch rats, a bounty of ten cents was paid for killed rats, and dead rat receiving stations were set up around the city. People were instructed to use gloves to handle rats and to immediately drop dead rats in kerosene or boiling water to kill the fleas. San Franciscans were good at catching rats, and the bounty for rats was so successful that it had to be cut in half.

All told, the San Francisco plague epidemic of 1906 was a completely different plague epidemic. Rupert Blue was lauded as a modern Pied Piper, and his success eventually led to his appointment as surgeon general (he attempted, unsuccessfully, to design a national health care program). He had his tricks while fighting the plague: to avoid the scrutiny of San Francisco businessmen, he wired Washington in code: to indicate the word *plague* he used *bumpkin;* to indicate *city board of health* he used *burlesque.* When the plague control campaign was finally over, Blue was honored at a banquet held in the streets. The theme of the banquet was "San Francisco is so clean a meal can be eaten in its streets."

THANKS TO THE PARANOIA OF politicians and businessmen during the first plague epidemic, plague had already spread from San Francisco into the rodents of California and the surrounding states, where plague remains today: there are more rodents currently infected with the plague in North America than there were in Europe at the time of the Black Death, though the modern rodents infected (prairie dogs, for example) tend to live in areas less populated by humans, as opposed to the rodents infected at the time of the Black Death. Kinyoun left the Marine Hospital Service shortly after his humiliation in San Francisco; he subsequently retired and moved to Washington, D.C., where he was working in the city health department when he died on February 14, 1919. Up until he arrived in San Francisco, he had had a brilliant career, and he had even spent time working in New York, where, in 1887, he had set up the Marine Hospital Service's first hygienic laboratory on Staten Island. There, he applied the principles of microbiology to cholera, and as a result, the first use of cholera culture was used to fight the entry of the disease through the port in 1898. Prior to this breakthrough, cholera epidemics had wiped out thousands of New Yorkers throughout the nineteenth century, and before being identified as a disease of poor sanitary conditions, cholera was blamed, first, on Irish immigrants, and then, at the turn of the century, on Jewish immigrants. Ironically, after helping fight this immigrant epidemic, Kinyoun spent the rest of his life complaining bitterly about his experiences in San Francisco—the eminent bacteriologist resorted to anti-Chinese and anti-Semitic slurs.

*Chapter 17*

# CATCHING

After the World Trade Center crumbled, the routine of the city worked like a tonic for my worries; in my view, the activities of a neighborhood corner were an opportunity for reassuringly cheerful monotony that caused me to marvel perhaps unduly, to rhapsodize, on the range of traits that man comprises. I do not know much about Freud, but I have a feeling that if a rat is indeed symbolic of a fear, then trapping a rat is, in a Freudian sense, confronting that fear, or at least putting that fear in a cage, so that when I went trapping again, it was therapeutic as well. On my second attempt to trap rats, I went with some people I know over in the city health department. The outing turned out to be historic: the city was trapping rats alive for the first time in several decades.

The group hoping to trap rats included Dan Markowski and Anne Li. Dan, the Tennessee-born vector control officer who had worked on World Trade Center rat control, was wearing a health department windbreaker, and his ponytail stuck out from underneath his cowboy hat. Anne was also dressed in a health department windbreaker and jeans. Born in Brooklyn, Anne is tall, with a dry sense of humor, and as far as rat experts go, she has a complete nonaversion to rats. She is an epidemiologist with the health department. Most of her work with rats occurs in a lab. Like me, she had never trapped a *Rattus norvegicus* on the streets. Anne had a lot of complimentary things to say about rats, such as "I think rats are so underappreciated." At another point, she turned to me and said, "Rats are the smartest creatures."

We were picked up by Isaac Ruiz, an exterminator who works out of the Lower East Side extermination office. He lives in the Bronx. Issac, who was wearing a wool shirt and sunglasses, told me that, as opposed to Dan and Anne, he was not especially eager to see rats. He is used to laying out poison for rats when they are not around.

We were all in a van going rat trapping for two reasons. First, the health department wanted an indication of how well their own rodent control measures were working; at that time, Bushwick was a test area for the city's rodent control program. The other reason was a result of the fear-rich postdisaster conditions that the city endured after the World Trade Center came down. The health department was trapping in Bushwick because of plague. At the time, the health department was working with the Centers for Disease Control and Prevention, which was interested in knowing about the rat populations of several major cities and how those rat populations would react if they were infected with plague. The CDC was especially interested in rats after the World Trade Center was destroyed and, a short time later, after anthrax, a biological weapon, was sent through the mail. What if someone attempted to bring plague to the city? How would the rats react? How should New York rats handle rat-infected fleas? So Dan and Anne were out in Bushwick, practicing before the arrival of the federal biologist, a rat trapping dress rehearsal, a little homework inspired to quell governmental concerns about the possibility of New York City hosting the Black Death.

And so it was on a crisp, clear morning that we pulled out of the City Hall area and drove through Chinatown and into the Lower East Side and then got stuck in traffic until at last we climbed up to the crest of the Williamsburg Bridge, where, in a brief fermata-like moment that involved a lot of neck craning, we could see from the tall, towerlike housing projects of lower Manhattan with the Chrysler Building and the Empire State Building behind them, over into the quilt of tenement buildings and low industrial operations that characterized our destination, the Bushwick section of Brooklyn, where I had previously (as described above) been with the city's rodent control department as they attended to a young girl bitten by a rat. Into the wilds of Brooklyn!

* * *

Bushwick—first settled by the Dutch, who, as one translator has it, called the area *Boswijk*, meaning "heavy woods," which probably were heavy until the woods quickly filled up with Germans, who had moved across the East River from their crowded German neighborhoods on the Lower East Side. The Germans opened up breweries and, in the mid-1800s, made Bushwick the beer capital of New York, at a time when men, women, *and* children drank an average of two barrels of beer or ale a year. It was subsequently settled by English, Irish, Russian, Polish, Italian, African-American, and Puerto Rican immigrants, and—after some of those people moved out—by people from the Dominican Republic, Guyana, Jamaica, Ecuador, India, Korea, and Southeast Asia. Bushwick was once filled with textile factories and textile workers, with breweries and brewery workers, but it was nearly destroyed in 1977. That was the year there of the blackout, a New York-wide power outage, and in Bushwick, after heavy looting—its main thoroughfare, Broadway, burned down almost completely—40 percent of the businesses closed up within the year. Still today, it is one of the poorest neighborhoods in New York, a place filled with abandoned lots, a place where 40 percent of the population is on government-assisted programs but where artists have just recently begun to sniff around and smell (relatively) cheap rent and lofts and other artists, a place where the city has very slowly begun to build subsidized housing.

In Bushwick, our van stopped alongside an abandoned lot underneath an elevated subway train. As the train thundered overhead, the light flickered on the street below: a children's flip-book scene of Myrtle Avenue would show the green and seemingly grayish green of a vacant lot filled with construction rubble and long-weathered paper trash, with crabgrass, dandelions, and thorny vines, with the raggedy-leaved mugwort that is a city relative of Western sagebrush. It was an abandoned lot of urban America, the outer-borough habitat of the North American wild rat.

In a few minutes, we were laying out traps. Dan joked that he was not permitted to describe the bait he was using: "We can't tell you the ingredients," Dan said.

I looked over him and he smiled. "It's peanut butter," he said.

"All you need is enough for them to smell," Anne said.

"Just a little dab'll do ya," Dan sang.

Anne went on at knowledgeable length about rat populations: "The general consensus," Anne said, "is that if you see one, then there are ten, and if you see them during the day, then you don't know what you've got." She talked about the behavior of groups of rats, about how rats share their stress within a rat colony, how they pass it on, like a cold; when one rats enters a colony and is stressed for some reason—because of aggressive behavior it faces from a larger rat, because of a lack of food—the rest of the colony will soon be stressed too. This group behavior is thought to be regulated by pheromones, substances that when secreted influence the behavior of other animals of the same species. Anne said that this kind of behavior regulation is often true for humans as well, and she mentioned a study that one of her graduate school professors had conducted among rats, and then among women. The human experiment, conducted among a group of young women living in a Chicago dormitory, showed that pheromones helped regulate the menstrual cycles amongst the group. One of the women in the dorm was a future senator for Bushwick and for all of New York State, Hillary Clinton.

Anne motioned across the lot, to where she had just placed a trap, and said, "Frankly, one of those vines had rat piss on it, which will help. You know, some people ask, 'Do you have to use new traps?' And that's the stupidest thing I've ever heard. Even stressed-rat piss is good."

I watched them lay down traps for a while, looking for a secret or trick. They were merely watching for possible rat activity, for obvious along-the-wall corridors, for cozy rat places.

Issac also laid out a couple of traps. He was a little tentative. When he got toward the back of the lot, a broken fence opened on the backyard of an old tenement building. A dog was barking. "That sucker gets loose and I know I'm dead," said Isaac.

Two men walked by and shouted out, "What are you doing?"

We told them we were trapping rats.

"The whole place is full of rats," the first man said.

"I catch 'em behind the fridge," the other man said.

"We want them alive," Dan said.

"They're hard to catch alive," the first man said.

THE NEXT LOT WAS AN abandoned corner property, on a back street in Bushwick, adjacent to a caved-in apartment building, attached to a little roofless wooden garage, which contained a collection of car parts, construction debris, and household trash: old speakers and a baby stroller jutted out of the pile; malt liquor bottles floated on the top like buoys.

"This is a perfect spot for death and pestilence," Anne said. They set down ten traps.

The other lot that they trapped in was along Grove Street below the elevated subway line. It was a small, fenced-off triangle of sandy dirt, the locked fence maintained by a community group a few buildings down Grove Street. The community group, called Make the Road by Walking, brings in services to the neighborhood, stops sweatshops, and often attempts to help rid people or streets of rats.

We all walked down to the community center, where we saw photographs of people in Make the Road by Walking bringing a rat to City Hall—a dead rat nested on a plate full of lettuce. Make the Road by Walking only had one key for the lot, so Anne and I went out to get a copy made. We walked along Myrtle and down to an intersection that was full of little stores. I could smell barbecued chicken and peeked in a window at a flock of roasted chickens and a sign: POLLO A LA BRASA. Some of the health department rodent-control field-workers say that a severe rat infestation depends on at least one good chicken place in a neighborhood; people buy chicken, take it out, and leave trails of chewed wings and bits of breasts.

In the hardware store, poison rat baits and traps for killing rats surrounded the cash register. The man at the counter said he had an excellent business in rat-killing products, a good sign for rat trappers, a bad sign for the neighborhood.

Back at the corner we entered the lot. I took a trap and was thinking about where to put it. So much of the lot seemed like good rat-trapping territory. The dirt there was perfect for burrowing, and rat holes were all along the fence where the abandoned lot bordered the sidewalk. Rat-wise, it was as the Great Plains must have been before white settlers killed all the buffalo.

As Isaac observed me, I looked for a place to lay my own trap—and finally chose a spot along the fence, near a burrow, a rat-traveled and rat-hopeful spot. I set my trap.

At some point, a rat ran out of a burrow and across the lot and down into a hole—a gray blur.

"Look!" somebody said.

Spotting a rat in the daylight made Dan confident. "It's only a matter of time," he said.

We locked the lot back up. The plan was to come back the next morning to see if we'd caught anything—it wasn't as exciting as the first time I went ratting, but this method seemed likely to produce better results. Isaac was going to drive us back to the city, after stopping in Brooklyn at a brand-new Russian bagel place for lunch. But just before we drove off, two men walked by and stopped at the fence; they looked into the abandoned lot and spoke with Isaac in Spanish. They told Isaac that they remembered when the lot was the site of an old wooden house that had become abandoned and filled with rats. They remembered the house being demolished and partially buried—the basement was still there, they said. They pointed to the ground, saying that the old home was still beneath it, still rat-infested.

The older of the two men shook his head. *"Las ratas, son el pan nuestro de cada dia,"* the man said. "The rats are part of everyday life."

AND THEN, AT LONG LAST, in a thrilling finale the very next morning, I caught a rat. What a heady feeling it was, to catch a rat! What a thrill to have briefly arrested one grimy twitch of the city's energy, to have isolated a note of the gray rat masses—to look a rat in the eye (even if it wouldn't look at me) and see it as a fact so ripe, as a city truth and a biting truth at that! The rat was there in my cage when we pulled up and I jumped out of the van with everyone else. We couldn't get into the lot because it was still locked, but we could see the rats in the traps through the fence. We were like children looking at unopened presents.

When we found the key, we all went to the traps that we had set, all of which held rats. The rats were testing the steel, pushing at the doors, then recoiling into the corners. The rats were not making any noise but they could be frantic. "These guys are mental in here!" Isaac said. Some of them had pulled in bits of plastic trash that they had shredded to create makeshift nests in the corners of their cages.

Anne picked up her trap and looked at her rat. She inspected it closely. "Look at this rat. This rat is beautiful!" she said.

I have to say that I felt my rat was one of the biggest, though I admit Isaac's could have been bigger. Hard to say, given how quickly they were moving back and forth in the cage. Especially in the setting—a scraggly lot, edges littered with garbage—these rats were the platonic form of wild.

Isaac looked at my rat and said, "It's pretty good, you know. But I went for the bigger hole." He winked. "I figure, the bigger the hole, the bigger the rat," he said.

We brought the cages from the edges of the lot into the center, where we took photos of the rats with the humans and then photos just of the rats. We placed the cages near each other and compared them. We covered each cage with a garbage bag to calm the rats and put the cages in the back of the van. Before my rat was packed into the van, I took a long look at him. The gray back fur on top, the lighter-and-almost-white fur below. The long, rat-yellow teeth, the startlingly dexterous pink paws with their long, thin, pink and prune-textured digits.

*A rat!*

We got in the van. As we drove to the next lot, we could hear the rats jumping around in the back, which didn't seem to bother anyone but me.

"I can't get over how lovely they look," Anne said.

"Yup, very nice," Dan said. "And this is like the size I was expecting."

"They're beautiful," Anne said.

At the next lot, beneath the M train, we all marched into the field. "You got one here," Isaac said. "Dan got one." The other traps were empty except for one. One of the traps had caught a starling. The starling had apparently flown into the trap and attempted to eat the peanut butter. Dan released the starling.*

"What is the matter with this lot?" Anne asked. As the M train ran overhead, a mangy old cat appeared in the lot. "It can't be that old cat," Anne said.

Dan shrugged his shoulders as he used his knife to scrape peanut butter from the traps before packing them into the van.

IN MEETINGS BACK AT THE health department, the rat trapping team had debated releasing the rats after catching them. On the one hand, they didn't want to tamper with the rat population for the sake of the

* European starlings were introduced to America by a New Yorker, Eugene Schieffelin, in Central Park in 1890. Schieffelin was the chairman of the American Acclimatization Society, a group of scientists and naturalists that sought to introduce animal species to North America. In 1864, they released English sparrows in Central Park and also introduced, or attempted to introduce, Japanese finches, Java sparrows, English blackbirds, and the English titmouse, among many others. They corresponded with other acclimatization societies, such as the Cincinnati society, which successfully introduced the skylark in Ohio. The society was also interested in introducing American fish to European rivers. Introducing starlings in Central Park was only a part of Schieffelin's plan to introduce to North America all of the birds mentioned in the works of Shakespeare. *Henry IV*, part 1, mentions starlings: "Nay, I'll have a starling shall be taught to speak nothing but 'Mortimer.'" The eighty starlings multiplied quickly. It is thought to be the most numerous bird in the United States. They are known to spread spores of toxic fungus, and they have contributed to the decline of the Eastern bluebird. In 1973 and 1974, in Kentucky, the governor called a state of emergency as swarms of them blackened the sky; at Fort Campbell, then the base of the Army's 101st Airborne Division, the birds threatened the helicopters. Their scientific name is *Sturnus vulgaris*.

experiment. On the other hand, they didn't feel it was appropriate for the health department to be releasing rats into the community—the people of Bushwick would not like it. Therefore, the plan was to check the rats for parasites and then to study their blood. For a blood sample to be effective, the rats would have to be alive when the blood was drawn. As part of the dry run for the visit from the biologist from the Centers for Disease Control the following week, Dan was planning on drawing some rat blood for testing. Now, Dan looked for an out-of-the-way place to draw blood from the rats. He chose an abandoned shack next to the abandoned building.

Isaac brought the rats out of the van, then went back to the van and waited alongside it. The idea of watching Dan draw rat blood repulsed him. Anne, on the other hand, was eager to observe; she hoped to learn how to draw rat blood.

As they were setting up, Dan stopped for a moment to answer his cell phone, a call from people back at the health department. Hanging up, Dan said, "This is the thing people don't understand. It's hit-or-miss science. At the meetings, people are like, 'Have you set a goal for the number of rats you're going to catch?' And the answer is, we'll set the traps and we'll get what we get."

On the floor of the abandoned shed, they cleared away what appeared to be the debris of a crack addict and brought out several clean syringes, some blood containers, cotton swabs, and a bottle of halothane, an anesthetic gas. The wind was blowing hard; it was slamming a door on the abandoned shed, which repeatedly startled me. The scene seemed illicit somehow.

As we prepared to look closely at the rats, Dan cautioned me not to make too much of them; he seemed to be saying that I shouldn't get caught up in rat lore and rat mystique. They were only rats, he explained to me, easily sedated, easily worked with, even if they were wild. As he spoke, he opened the garbage bag covering one of the cages. Ann treated two cotton balls with halothane. She dropped the cotton balls into the garbage bag and twisted it shut tight, to put the rat to sleep.

In a few minutes Dan looked into the garbage bag at the rat. He shook his head as he closed the bag and looked a little incredulous: "He's livelier than he was before."

They increased the dosage of anesthetic, putting in three treated cotton swabs this time. The wind was picking up. They waited and looked in the bag again. The rat was still alert. Whereas Dan had begun this work by cautioning me to remember that a rat was nothing more than a rat, now his feelings about the rat seemed to have changed. "That rat's one tough bastard," Dan said.

Dan increased the dosage again.

Finally, the rat looked unconscious, its tail limp, though when Dan took it out of the cage, he quickly discovered that it was still awake. He held the rat down on the ground with his hands and placed a halothane-treated swab directly over the rat's nose, holding the cotton with tweezers. The rat was going from groggy to woozy to sleepy to asleep, and as this happened, I realized that the rat was a large female, measuring, as we later determined, about eleven inches long, not including the tail, which was close to another ten inches long and looked to me like something off an armadillo. At last the rat seemed at peace. Dan held the rat down on the ground and plunged the needle into its lightish chest fur, aiming for the rat's heart. He drew out the rat blood and bottled it; the blood was a deep and rich red and mammalian, the color of human blood. As this happened, I looked away, over at Isaac, who was still standing next to the van. Dan put the rat in a Ziploc freezer bag, with another dose of halothane. The halothane would kill the rat: from sleep to death.

On the second rat, they started with a larger dosage of halothane. The second rat was larger, a foot long. "He's *very* healthy," Anne said. "He's missing a little hair on his nose but I think that's because he was trying to get out of the cage."

"Anne, you want to bleed this one?" Dan asked.

"Okay," Anne said.

Two cotton balls soaked with halothane were not enough. They tried more. "This is what, four balls?" Dan asked. "Man, four balls after taking two balls already." Dan shook his head again and held a cotton ball to the rat's big nose. "We're underestimating these rats," Dan said.

Anne agreed. "These guys are fucking amazing."

She began to draw blood from the rat, but as she did it became clear that the rat was not asleep.

"Uh, he's still awake," Anne said.

Dan had already noticed this. He had seen the rat moving, seen it reviving and regaining consciousness, and it seemed to me that, rather than wasting precious rat-crisis time alerting us to the rat's movements, he was deep into securing the small abandoned lot, the rat-blood-drawing site, and in an impressively unfrantic way looking for a means to stop the rat, any way to secure the rat, in fact—so that finally he just stepped on the rat's tail. "Oh, you were just play asleep, weren't you?" Dan said. The rat rose up and seemed to slash at Dan; it snarled. In a minute, they had it anesthetized again.

The wind blew and the shed door slammed and I saw a white cat come out of the trash, emerging from behind a plastic replica of a Greek bust. In knocking out the third rat, Dan and Anne increased the halothane dosage significantly. When they thought the rat was asleep, they took it out of its cage. It too was a large, healthy rat, a foot long. Dan was beginning to bleed the rat. As he did, he happily noticed something on the rat's fur.

"What's that?" Dan said.

"It's a tick," Anne said.

Dan leaned into the rat. "It's a mite."

"Okay, Dan," Anne said. But as they were talking, the rat began to squiggle greatly.

This time, the rat's movement was more than just the groggy squiggling that the other rats had made; this time, even after a large dose of anesthetic, the rat was somehow recovering all its rat strength. Dan stopped drawing blood. He moved calmly but rapidly. He seemed to think about trying more halothane on the rat. Then the rat moved again—this time less *rat on drugs* and more *wild rat.* Dan put his foot down on the tail.

"Oh, that's it," Anne said. She backed away. Dan pulled up his foot. I was standing there too, right next to the rat, and I looked around to figure out where I could run, if necessary, then realized that I was between the rat and the abandoned house, trapped. So I just stood as still as possible as the rat lifted his big body up and first waddled, then walked, then wearily scampered off. I was trembling a little.

Dan was not trembling but he did seem as if he were in shock. "I'm using my foot," he said, speaking in the present tense, as if watching a replay, "but the rat *still* gets up and pulls himself away." Dan was really shaking his head now. In fact, he was no longer telling me to keep rats in perspective. "Rats are incredible, they *really* are," he said. "I mean, there aren't many animals you can bleed like that and that can take it."

"And the thing is, you're never going to beat them," Anne added. "They have something going for them that you don't, which is natural instinct. Like you don't *instinctively* get rats. You're not a cat. But they *instinctively* avoid death and the obstacles that we put in their way to kill them. Besides, I always say that if you killed every rat in New York City, you would have created new housing for sixty million rats."

"Let's put it like this," Dan said. "If you put that cat in that bag with the halothane, he'd be dead."

We all stood there and looked over at where the rat had returned to the wild.

"It's a New York City rat," Dan said.

PLAGUE HAS THREATENED NEW YORK CITY on several occasions. The first time was in 1899 when the SS *J. W. Taylor,* a British ship, traveled to New York from Santos in Brazil, where plague had broken out a few days before it sailed. The steward died on the boat and two men were sick when it arrived in port. The ship was quarantined. It was permitted to take on coal in New York Harbor, but the coal ship was ordered to stay fifteen feet away, to avoid the transfer of rats. A few days later, the ship sailed away, but a controversy erupted that fall when it was discovered that a shipment of coffee from the ship had made it ashore. (When asked if the coffee was ever drunk, the owner of the coffee company said, "I should hope so.") Once in 1948, the port discovered plague fleas on board the *Wyoming,* a ship traveling from a plague port in Morocco, but then the city secretly trapped the rats on the surrounding piers and in the surrounding neighborhoods and determined that plague had never gotten off the ship and out into the city. For decades, the port of New York constantly trapped rats and checked them for plague and plague fleas. They also fumigated incoming ships. Teams of exterminators were stationed out at Hoffman Island, a small man-made island that was built just off the shore of Staten Island in 1866, after rioters protested a quarantine station on Staten Island itself. In 1928, port exterminators were estimated to kill an average of ten thousand rats on ships every year. By 1952, a team of 14 men was exterminating rats, including Louis A. Lindecop, the port's chief sanitary inspector at the time. Lindecop once told a reporter he had seen rats go crazy for raw cucumbers but eat ship's grease if there was nothing else.

The Centers for Disease Control's concern with plague in New York City has to do with the use of plague as a biological weapon, which stems in part from the Japanese's use of the disease as a biological weapon during World War II. At that time, Shiro Ishii, a general and a doctor, led a special biological warfare research unit, called Unit 731. Unit 731 worked in Manchuria, where outbreaks of plague had occurred in 1910, 1920, and 1927. The general was interested in plague as a weapon because of its ability to create casualties out of proportion with the amount of bacteria necessary to disseminate the disease. Also, plague could be used militarily in such a way as to make it appear like a natural-occurring outbreak. Initially, General Ishii's unit had difficulty devising a way to drop plague bacillus from airplanes; the bacillus died by the time it hit the ground because of air pressure and high temperatures. He next infected human fleas with the plague, in hopes that they would infect humans and even rats, so as to prolong the epidemic. He attempted to spray fleas from compressed-air containers, which was not successful. He eventually built clay bombs and filled them with infected fleas and dropped them, which worked. Eighty percent of the fleas survived. He experimented on humans, and in these experiments, Ishii determined that if ten healthy people are in a room infested with twenty plague-bearing fleas per square meter, four will die of plague. (Anthrax is more likely to kill someone, but the plague will infect more people.) It is thought that the Japanese used plague as a weapon several times in China during World War II. After a plane flew over Changde, a city in the Hunan province, people began dying of the plague. One of the clues that has led plague experts to believe that the outbreaks were caused by humans was that the rats only began dying in the city two months after humans began dying. After the war, Unit 731's human experiments were made public; members of the unit had practiced vivisection on humans. General Ishii was never tried for war crimes. On the contrary, a deal was made in which he donated his records to the United States government—Ishii's records included fifteen thousand slides of specimens from approximately five hundred human cases of diseases that had been caused by biological weapons, including the plague and anthrax. He retired as a respected medical man.

The Soviet Union designed its biological weapons program using General Ishii's work as a template, and Americans incorporated Ishii's work into theirs as well, when the United States began experimenting with biological weapons in the 1950s. In addition to testing weapons to be used against military enemies, the Americans tested biological weapons that could be used against the United States. To simulate the spread of anthrax over large populations, the government used microbes that were similar to anthrax but thought to be harmless—*Serratia marcescens* and *Bacillus globigii*. In April 1950, two navy ships sprayed the residents of the Virginia coastal communities of Norfolk, Hampton, and Newport News with *Bacillus globigii*. Neither the public nor Congress knew about it. Similar sprayings were made in San Francisco Bay—area residents were exposed to clouds of these microbes—and in perhaps as many as two hundred other places around the United States. In New York City, the military released *Bacillus globigii* on the subways. In the summer of 1966, plainclothes soldiers dropped *Bacillus globigii*–filled lightbulbs on city subway tracks; they dropped them on the tracks in between the subway cars so that the wind of the train would whip the bacillus up and spread it through the system. Later, other soldiers carried suitcases with air samplers to see how far through the subway system the *Bacillus globigii* had spread. The results are classified.

The most recent plague case in New York City was reported in 2002, and it came not by boat nor by bomb but by plane. The plague in New York was the same plague bacillus that had come to San Francisco via China in 1900 and then spread out into the rest of the country before it was properly contained. Plague cases occur sporadically every year in America—it is labeled a reemerging infectious disease by the World Health Organization—but there had not been a human plague case in the city limits since the sailors from Brazil arrived in New York harbor in 1899. In November 2002, two tourists went to the Plaza Hotel for dinner and came down with flulike symptoms the next day. They were a husband and wife from New Mexico, and initially they thought they had simply overdone it with wine and a late night after a long flight; they stayed in their hotel room for twenty-four hours, but only felt worse and on the following day checked into a hospital. The doctor did not immediately recognize plague; the couple suggested they might have it since they lived in an area in New Mexico where several neighbors had contracted plague. They were quickly quarantined.

Because the couple was from New Mexico, where plague cases are not unusual, doctors quickly surmised that they had been infected with plague there, just before coming to New York—both mice and pack rats on the couple's five-acre ranch had tested positive for plague. They were immediately treated with antibiotics. The woman improved rapidly, but her husband's condition worsened: his kidneys appeared to shut down, he lost circulation in his feet, he went into a coma. In the end, he couldn't breathe on his own for three months. An avid outdoorsman, the man awoke to find that, due to tissue damage, doctors had amputated both of his feet.

When the couple first came down with the plague, the television news stations immediately mentioned the medieval pandemic and questioned whether the couple might possibly be terrorists attempting to smuggle in plague as a biological weapon to be used against the city. One of the stations used a dramatically lettered banner over the news anchor's head on the TV screen that said BLACK DEATH. The coverage of the couple leaving town, on the other hand, was low-key. When they were both well enough to return home, the man told people he felt lucky and vowed to learn to walk again. The woman said that she had kept a journal all through the ordeal as she sat by her husband's side. After she left, she wrote in her journal that she cried when saying good-bye to all the people in New York who were so nice to her. "Good-bye to this fine city," she wrote.

IF YERSIN, THE MAN WHO solved the medieval secret of the plague in Hong Kong in 1894, has a counterpart today, he might be Rusty Enscore, the biologist with the federal government who came to go rat trapping with Dan Markowski. Rusty is a commander with the U.S. Public Health Service assigned to the Centers for Disease Control. Rusty doesn't look like a guy who gets a call from the government during a possible bioterrorist event telling him to sit by the phone and be ready to leave at a moment's notice to handle an outbreak of an infectious disease—plague, for instance. He looks like a guy on his way to work at the office on a Saturday and then maybe fishing—though, to be certain, he is a kind of fisher of plague and thus knows it intimately, knows its details, its lore, its habitat.

He is about five feet ten inches with close-cropped hair, and a big grin, and when he showed up at the health department's offices to trap rats, on an unseasonably warm day downtown, he was wearing jeans and a T-shirt and carrying a briefcase. Rusty lives in Colorado, where he works for the Centers for Disease Control's division of vector-borne infectious diseases. Usually, he drives around the western United States with a portable lab attached to the back of a small recreational vehicle. He traps and tests and bleeds animals everywhere. He himself marvels at the variety of his itinerary. "I was in Denver trapping beavers the other day, and next Monday I'm going to be in Nevada bleeding sheep. Now, I'm in New York City trapping rats," he said.

Rusty was in New York for the first time. He was staying near Times Square but he hadn't had a chance to look around; he was picked up early by the rodent control van. Isaac was with Dan again. (Anne was on another job.) We all drove out near John F. Kennedy International Airport, giving Rusty his first view of the Brooklyn-Queens Expressway, which looked jammed with traffic to a nonnative but looked pretty good to Isaac—he was making decent time. The reason for rat trapping in the area of JFK was that the airport is thought of as a likely spot for someone bringing an infectious agent into the United States in an attempt to disperse such an agent. The neighborhood that we trapped in was part New York City, part neighboring county, and, as a result, feels a little forgotten by both—it only had the beginnings of a sewer system installed in the early 1990s.

Rusty explained his job to me in the van on the way. He talked about the reasons that the public health officers might want to know about a city's rat population: "When I land in a bioterror situation, the first thing I have to do is figure out the rodent population. If I already have knowledge that the flea infestation is low, then I can start killing rats right away. And that's what this is all about. When it happens, if it happens, how do we tailor that response?"

Rusty ticked off a list of other diseases that rats could be checked for: "There's bartonella, West Nile virus, plague, hantavirus, tularemia." Of plague in the U.S., he said, "What we don't know and what I'd like to know is why there aren't any cases in California right now."

And as we were driving, just for the heck of it, I asked him whether he had any doubts about whether the Black Death of the Middle Ages was caused by plague-infected fleas via rats—I mean, how often are you riding around New York City with a plague expert whom you can bounce plague theories off of? Did he subscribe, in other words, to the theory that anthrax had caused the Black Death? Not only didn't he have any doubts about the plague in medieval Europe, but he had studied Roman granary and taxation records (Romans paid taxes in grain—a percentage of yield) and seen indications of large plague-inducing rodent population increases as a result of bountiful harvests in the time of the Justinian plague pandemic, and he felt that a lot of extra grain was good evidence of extra rats.

We got out of the van on a run-down, swampy street, only partly paved, in the reed-filled borderlands of New York. In a place I'd never been, in the city of my birth, I stood there thinking about the city: *How ultimately unknowable it is, like a vast forest, and how different every little portion is, even though people who don't know the city (and even some people who do) think it is all the same, all monotonously tenemented and office-towered, like the set of an action movie!*

Meanwhile, everyone else had started looking for rat traps, the ones that Rusty, Dan, and Isaac had laid out the day before. We went into a falling-down, one-story house, surrounded by phragmites and garbage. Everyone was cautious because the day before they had walked into a house that they thought was abandoned but turned out not to be, and the Vietnam veteran they'd woken up wasn't happy about it. They checked the traps. Nothing. They went on to the next traps. Nothing. Over to some traps in the reeds. Again, nothing. In all, twenty-five traps were empty. Depressed, Isaac loaded all the empty traps into the van.

Everyone was pretty frustrated, especially after catching so many rats the week before. While we were waiting, Dan got down on the ground and looked through a bunch of old abandoned tires for mosquito larvae—it was getting on to the next infectious disease season, the season for West Nile virus. He dipped into the dirty water.

"You dipping for larvae?" Rusty asked.

"Yeah, sure." Dan's voice had a tinge of disappointment.

Rusty laughed.

"I mean, might as well get something done on this trip," Dan said.

A plane, its landing gear ready for the runway at JFK, flew over, trailing a deafening roar. It was a supersonic Concorde, a plane that was about to be decommissioned, an endangered jet species.

For most people, a lack of rats would surely bring a positive reaction, but the lack of rats was disappointing for the rat trapping team; it meant there would be fewer rats to check for fleas. Dan, as a host, wanted his guest to be pleased or even feel at home by catching a rat. Personally, I felt a little relieved by the complications that would obviously be involved with infesting rodent populations for sinister purposes, at least in this part of Queens—rats are harder to catch than you'd think. Ratless, we all got in the car and headed back to downtown Manhattan for a consolation event. We went to an old Department of Health building on First Avenue. The corridors were out of the fifties, straight and outdatedly antiseptic looking. Dan took out the Brooklyn rats and about a half dozen rats that he had subsequently caught in lower Manhattan. The rats were frozen. Each was in a Ziploc freezer bag, and they hit the lab counter like gray rocks. Then Dan and Rusty pulled out two microscopes and began combing the rats for bugs—specifically, for Oriental rat fleas, the flea with the best track record for transmitting the plague. They were looking in the microscope for the absence of a telltale black band—which would indicate the presence of New York's version of the bug that nearly wiped out civilization.

Now the two men were stooped over their rats, concentrating, making idle, parasite-related chitchat, talking about fleas in the windowless room, under the fluorescent lights that tinged human skin a sickly green.

"They'll frequently go to the eyes to drink, and a lot of times they'll go to the crest of the back, simply because the animal can't scratch there, but there's no one particular place," Rusty said, not looking up.

"That's the good thing about ticks, they will go to the head and neck area," Dan said.

"Ticks, in my experience, seem to have some sense of gravity because they head up."

"Absolutely," Dan said.

"I'll take a tick if you don't want it, but I hate mites," Rusty said. "Mites take a special mentality."

"Oh, yeah," Dan said. "They are just so small and so hard to see—what a nightmare!"

They went through a few Brooklyn rats, noting again their size and health. Then they started combing fleas out of the Manhattan rats—rats caught in the vicinity of Wall Street and City Hall. They did not find any Oriental rat fleas, indicating that plague might have some difficulty taking hold in New York. However, they did notice that the Wall Street rats were especially stressed looking, beat-up, sickly in comparison to their Brooklyn counterparts, an indication of increased disease risk.

"Look," Rusty said, "when you comb back the hair on this one, you can see the bite marks on his ass."

"Oh, yeah, where's that one from?" Dan asked.

"Lower Manhattan," Rusty said.

"See that?" Dan said.

*Chapter 18*

# RAT KING

ON WINTER MORNINGS, I awoke groggy from my rat duties, kissed my family good day, then filled my thermos and subwayed through my day in the shoulder-crunching, welcoming, belligerent, and ambivalent crowds of Manhattan and Brooklyn and, once in a while, Queens, Staten Island, and the Bronx. I was not delinquent in my rat duties. No, I undertook them with greater zest as I was desirous of arriving at some truth about the rats, or at least about my alley—some great notion that I felt still eluded me about my rats and my particular situation. I didn't know what that truth was exactly, so I took a more scrupulous view, despite the cold, because in my mind, or in my sleep deprivation, I felt close to it.

## FROM MY RAT JOURNAL

5:04—Dusk. A small telephone company construction site, a hole burrowed by phone company workers, in the center of the alley, surrounded by cobblestones stacked up like dentures, surrounded, in turn, by piles of dirt. Up at the end of the alley, a van from a catering place. Two men carefully loading it with what appears to be food.

5:09—The van, loaded, attempts to drive around the corner of the alley, to proceed from Ryders Alley into Edens Alley, but is thwarted by the angle of the alley's turn. Instead, the truck backs up, driving across plastic trash bags, crushing garbage—a liquid bursts from the overripe bags. A trickle of garbage juice, a stream is born.

5:19—A man in white pants, shirt, and apron emerges from out of the kitchen of the Chinese restaurant, lights a cigarette, relaxes amidst the refuse. Confucius taught that one eats and one does not know the savor of food. Likewise, one does not know the potential for rankness and repulsiveness of the once-savory food.

5:24—I am sipping coffee, settling in for a few hours of observation, when I see it, the first rat of the night. The rat appears at the top of the alley. The rat stops. The rat crosses the cobblestones, bounding, stopping once, bounding again. The rat circles behind the construction site and then comes back across the alley again to the garbage on the Irish bar and restaurant side. I try to maintain a certain rationality, or clinical aloofness, and yet (as is typical by now) I am mesmerized—first, by the appearance of a rat, still a perverse miracle to me, and second, by a movement that is completely ratlike in its hugging of the wall, in its bold carefulness, and yet also different—a surprise. Who could have predicted that the rat would cross the alley up high and then cross down low again? Seconds pass. One, then two more rats follow. Ah, the rats of Edens Alley!

5:33—A man emerges from the Irish bar, bringing out more garbage into the alley's growing stream: the sound of the door opening, the crackly thump of the heavy but light-plastic-enclosed trash causes rats to scurry out of human sight.

5:40—Another rat has appeared at the top of the alley, has squeezed himself up through a hole in the sidewalk, paws pulling him up, up and out. On the Chinese restaurant side, I see three adult rats and two juveniles—at least I think the juveniles are juveniles: they are in and out of the garbage bags quickly, with their youthful vigor, their close-to-the-street risks. Juveniles seem to wander farther down the alley, nearly entering the street; the older rats stay closer to the nests. In their small size is the promise of rat regeneration, of rat life reborn in this still-winter alley, in the wounded but healing city. And there seem to be more and more juveniles.

5:44—The rats retreat suddenly. The reason: three men enter the alley, though when I see the men, I wonder which creature left the alley for which creature—sometimes it seems as if the rats' departure is a courtesy extended by the rats. I leave my post at the bottom of the alley and go around the corner and, from out of sight of the young men, see rats in the garagelike lot on Gold Street; peering through the fence, I notice rats climbing over lumber, up metal scraps, across broken things. I think of all the rats that have crawled through this alley before, the history of this alley's previous inhabitants. Oh, to know—to *really* know—this pellicle of rat-infested ground.

I can hear the rats too: scurrying, screeching, their strong nails scraping the scrap construction metal. Still waiting for the young men, I walk up into Edens Alley and see more garbage and then more rats and then *more* rats coming around the corner from Ryders Alley—still being displaced by the three humans. I sneak around, quietly, but at some point I begin to believe that the rats are aware of my presence, and consequently, I back down from the alley, slowly, then more quickly, then a little more quickly still. My movement is noticed almost peripherally by a man who is at this time passing the entrance to Edens Alley, arm in arm with a woman. When he sees me seeing the rat, when he interprets my rat alley evacuation body language, he picks up his own pace. And like me, he nearly sprints, slowing down a few seconds later, just down Gold Street, when his partner looks at him strangely, at which point he says, excitedly, even slightly frantically, "Holy *shit!* Did you see those rats?"

5:55—I am back at the base of the alley, trying to be discreet, but the young men in alley are still there, making loud chattering noises. They notice me. Again, I scurry off. I retreat to John DeLury Plaza, reflexively conjuring up John DeLury himself for an instant: the pipe, the glasses, the stubborn, shout-prone, *un*-unrelenting negotiator, his workers. And then I move farther across the street to stand near the trash in front of a Burger King. The young males continue to peek out of the alley at me. Who knows what they are thinking when they are looking at me? Though I was concerned, I was also preoccupied. I had to stay with the rats, no matter what.

6:03—More garbage comes up out of the bottom of the Irish bar.

One bag lands on a rodent bait station that is ancient and nearly destroyed. The garbage tide is rising. I am reminded of Milton, in "Lycidas": ". . . tomorrow to fresh woods and pastures new." Though when I am reminded of it the words *woods* and *pastures* are replaced by *trash*.

6:08—The young men move out. When will the rats return? And, continuing on that line of silent inquiry, what exactly am I waiting here for? Nature, even rat nature, does not answer mortals, even rat-interested mortals. If the alley speaks, it is obscure: *Claude os et audit!*

6:14—Sixty-two people pass in thirty seconds; even at night, even after rush hour, even when fewer people are on the street because of the World Trade Center attack and because of some lingering fear and even still some panic. Even after all this, New York is stuffed with people, people constantly walking, running, going out to eat, leaving food behind, even if they don't know it. Extrapolating, I calculate that at this slow, semi-abandoned, semiconscious downtown rate, the entire city passes, all eight million New Yorkers, in a month and a half. To stand in an alley is to watch the city from its bowels, to feel life grumbling in its gut.

6:15—The first rat returns. He is large. He comes from out of the deep hole, that great hole in the back. I might plumb the depths of that hole, but to what end? Suddenly, as I ponder, the geographic, historical, and animal begin to collide. I think about the hole. I know, for instance, that this is the hole that leads to the basement of the sanitation workers' union hall—that is the raw, coincidence-flavored fact. And yet to my eye, it appears bottomless, as if it goes straight to the other side of the world. And so I wonder further, what *is* this hole, this deep pit?

6:18—They are all returning, the rats, flushing back out into the alley, like beachgoers after a brief thundershower. As I step slowly into the alley, I am now focused on the hole, the great rat pit. My focus is broken momentarily, though, when someone drops a cigarette from an upper floor of an apartment above the alley. The cigarette hits the ground: a miniature, sparkling orange meteorite lands at my feet, a break in the firmament. I look up and see a faint star.

★ ★ ★

It seems inevitable in retrospect. Like a fisherman who becomes fixated on a fish, like a whaler who becomes obsessed with one whale, I notice one rat: a large male, with an unusual tail, strangely curled. This rat appears, through binoculars, to be a little *more* than a foot—i.e., longer than a cobblestone, which is easily measured to be twelve inches. I watch this rat graze from one garbage bag to the next. This rat does not leave the alley with food; he eats food in the alley, standing his ground. And most significant to me, when it goes away, it goes into the hole.

I move to test this rat. I flinch. It flinches, but it does not flee upon my movement. It hunkers, moves back slightly, and in a few seconds it returns to a large garbage bag, where another more moderate-size rat pulls and pulls at garbage. The big rat joins in. There appears to be no animosity between the rats. They stand on their rear legs. They pull and pull, until first the large rat and then the moderate rat each draw from the bag a large piece of chicken. Again, they do not quarrel. With an abundance of garbage, there is a harmony in the rat alley.

6:32—A sanitation truck passes, and then a street cleaner. The rats are unfazed.

6:42—All at once, with a silence-breaking clatter, a small pack of men pour out of the back door of the gourmet supermarket at the end of the alley. They have bags and bags of garbage and they fill the large Dumpsters full of trash. They overfill them with garbage, using poles and gloves. The rats have retreated and the men from the market are now tossing trash down Edens Alley, fire brigade style.

6:50—Out across Fulton Street, the garbage from the Burger King is dragged out, as I imagine is happening at fast food restaurants all over New York at around this time. A small mountain of garbage bags forms, a vile and grease-dripping sedimentary New York City occurrence that nightly turns the streets into miniature badlands, to be eroded by morning, assuming the sanitation workers arrive, after which there will be dark stains on the concrete, like sweat on the morning rocks of a mountain.

6:57—More garbage, more rats, so many more that it's becoming difficult to concentrate: there are too many rats now, more than a dozen visible at any time—squads constantly surfacing, resurfacing. In the foreground are the young rats. In the back, the larger rats, the rats that must be older, given their size: when I venture up with binoculars I can see their mottled coats, the bite marks—on one, a gashlike scar. I see also specialty diversions, rat performers in a circus of trash, affording much entertainment for the alley watcher: a rat climbs up a garbage bag, stops at the summit, appears to look around. The rat jumps, nearly straight up, in fact jumps for what my later measurement will show to be one foot—up, up and onto the old ledge of a boarded-up window. The rat walks along the ledge and turns, behind the rusted old steel window bars, to face the alley again, then lowers himself down on a bag back close to the wall, a bag that is inaccessible from the alley floor.

7:15—The rats are drunk on food, I think. Technically speaking, all a rat needs is three or four ounces of food a day, but these rats seem to be greatly exceeding that amount, and wouldn't you? It is not at all difficult to picture the rat eating at its food source until the food source is destroyed, cleaned out, until the rat must move on to the next alley, the next street, the next neighborhood. Now, the rats that grab food and run back to the nest are getting food and running around in circles—as if feigning a return to their nests, or maybe not feigning, I couldn't say. In a few minutes, they are not eating as much, but seem to be recreating, playing. They fool in the little pile of dirt by the telephone company's excavation; they burrow, throw dirt, run off. How free they are! How full of liberty in what was not originally their own environment but is now! Perhaps this alley reminds them of their past in some distant way? In their voyage from garbage to nest, from nest to garbage, with the slight variations that come when they lead each other toward food with the scent of food-hunting success or away from danger with the invisible scent of stress. Do they *understand* old paths, old routes, old rat roads? Does this dirt-filled burrowing spot perhaps remind them, somewhere deep in their genetic structure, deep in their rat bones, of a place where they burrowed freely in

Siberia, in the rat-originating Eurasian steppes? Or of their first burrows in old New York?

7:25—I am moved to move by a man who, apparently not seeing me in the alley, was moved to urinate in the spot where I had been standing. I cross the street and sit in John DeLury Plaza on my portable camping stool. I drink coffee from my thermos and think once again of John DeLury and the time of trash piled high in the streets. I think of big rats in general and then the big rat in the alley, and then I look back in the alley and easily spot him and his corkscrew tail. Some people go off into the mountains to collect themselves and look into their souls, but here I am enjoying the view at something outside my soul, in this case a rat.

7:32—Another guy comes out of the Chinese restaurant. He is kicking boxes as he smokes a cigarette. He leans back when he kicks the boxes, keeps his body back, at a safe distance. Is he too observing rats? Or is he merely repulsed? In a big city, or in any city for that matter, it is one thing to observe someone who appears to be watching rats, and quite another to know how they might feel about them, especially when you yourself having been watching rats for three seasons, and you're still not certain precisely why it is that you are watching them.

IT HAD BEEN A MILD WINTER, to be sure. The following year, I would watch snow fill the alley and notice that the rats coming up from their burrows in Edens Alley tunneled through the soot-peppered ice. But during this winter, snow fell in the alley only lightly, and I wondered if the mildness had somehow helped explode the rat population because on another night, closer toward spring, around eleven o'clock, I saw even more rats, the rat-infested alley seeming more rat-infested. I counted eighteen easily but then lost track. To some extent, the alley seemed clean; it had rained recently and the light from the streetlight glimmered on the slick cobblestones, on the garbage bags' luminescent black. But the alley was more ratty. This evening, for whatever reason, things were not as amicable among the rats. The rats were squealing. The rats were fighting. Juveniles streamed down from the large dark hole on the south side of the alley and gorged in the Chinese restaurant's refuse. Adults settled into the trash by the door of the Irish restaurant and bar. At the foot of the alley, I heard a man passing by on Fulton Street, tossing off conversational litter: "That's what life's about—choices." And then I looked to see that in the alley, the rats were mating.

If I draw the line at something regarding rat observations, then it is rats mating. I prefer to let them mate in privacy, though I will say this: the male seemed aggressive, and the female made sounds that seemed to indicate she was not interested in mating, though her movements indicated interests to the contrary. But mate these rats did, and repeatedly, which is a thing about male rats—they have been known to mate with a female rat long after she is interested in mating, sometimes after the female is dead.

When they were done, I inspected the male rat again. His tail was distinctive. Was it the same corkscrew shape I had seen before, or was I just imagining it? Had I spent too many hours in the rat alley? Either way, I could see this rat chasing another rat. This rat was chasing a rat that had come down from the top of the alley, down the hill that begins at the deep black hole. The rat chase stopped at the end of a prescribed range, an invisible (to the nonrat) border that describes the difference between home and not home. The rats ran to the line and stopped, as if encountering an invisible fence. The chase could have been friendly for all I know, but it could have been to the death; rat territory seems sacred to rats.

A FEW NIGHTS LATER, MIDNIGHT, and the rats were in full swing, and I was looking at the rat that I was amazed to find I recognized. I waited a long time, because the bar around the corner was crowded, and the alley accepted the overflow. Tonight, a small group of young women were assembled in the alley, standing alongside the dark rat pit. "They're carding," a woman said, voice high-pitched. Two young males arrived. The young males laughed, then walked away. A woman said, "He ratted on us!" When the young people finally left, the rats returned to take their place. Picture me, in fleece and wind-resistant overcoat, furiously scribbling notes. Picture me looking up, amazed, because I saw the rat, the rat with the tail. Picture me understanding a little bit about this community of rats, recognizing some traits, some habits, some of the players in the colony—or at least recognizing what must have been the alpha male.

THE NIGHTS WERE WARMER, and a week passed, and on the next night I was in the alley, it was pouring rain, and more people were in the alley, this time a film crew, filming a scene: one man attacking another, a mugging. The actors were at the top of Edens Alley, and as I watched them stand where the rats normally skitter, I wondered what story, if any, the rats would tell about this gutterlike lane.

Two policemen in a police car were watching the movie set, the police car's headlights spotlighting the dancing of the rain. (Policemen were present because of a city law that requires police presence on a movie set on which any weapon, even fake, is used.) I waited on the edge and watched as the rats got used to the actors. As the rats emerged, they appeared in the police headlights like stars in an ice show. After a while, I introduced myself to the director as someone who was observing rats. Of course, I was happy to hear the director mention that he was pleased with the presence of rats—he said they would help the scene, which was about a robbery, as best I could understand. I did not mention the big rat, the one I recognized. This didn't seem like something I should mention to him, or to anyone, for that matter. This particular rat eventually emerged for its cameo, though, and when it did, the director turned to the cameraman, recognized the rat as the star that it so clearly was.

"Oh, man, did you see that rat?" the cameraman said. "Jesus, that is one *huge* rat!"

* * *

MY RAT, MY LEADER OF RATS, my rat that doesn't seem to so much lead as to coerce—my Rat King, which I called it even though I knew it was not a huge Rat King that sat on a ring of other rats' tails, that ruled other rats, as best I could tell. I saw him as a star among stars in the deep and capacious alley of rats. For me, this rat cast a transcendent sublimity that united these unwanted inhabitants of the alley in particular and the city in general, even if they are abhorred. I saw him as the Brute Neighbor in all of us, the representative Unrepresented Rat.

But was I just making him up? Was he a rat of my imagination?

A few nights later, on a night when the private trash-carting truck came to take the rats' habitat away, to open the truck's hydraulic jaw and engulf the trash, I was caught unawares as the truck arrived; I was startled. I backed off as the trash was taken away, as the driver of the truck climbed down into the alley. I was up against the wall when the driver approached, nodded, greeted me with no apparent malice; on the contrary, he was smiling. I said I was looking at rats. The driver didn't blink. "Did you see the big one with the tail?" he said.

I was flabbergasted.

I described the rat. I described the rat's tail. This man knew this rat's tail.

"Yeah, he lives back in that hole in there," the man said. "He's big, boy. I've seen him *walk* up stairs."

*Chapter 19*

# A GOLDEN HILL

THE RAT HOLE! The rat pit! The darkness that is home to my Rat King. Near the end of a year spent thinking about rats, after three and nearly four seasons examining this squalid parcel of city land, after communing with this Philosopher's Hole, if you will, I am lured by a rat to consider this aspect of the alley of the Rat King. I see the rat, in my mind's eye, climbing down into this hole—this fire-escape-equipped darkness that is, on flashlight-led inspection, more than one story deep. I see the hole. I walk down and out of the alley and around the corner and see the front of the building and realize that the building itself is built into the side of a hill: its back to the hill's incline, its face facing the slope down, which, in turn, explains the steep slope of the alley, which explains—at last!—why the hole down to the basement is so unusually deep. So now, when I follow the rat down into the hole, I am thinking, paradoxically, of the hole's topographical opposite; I am thinking of this hill, this hill that I never really noticed before—I am thinking of Larry Adams, the city's exterminator, who talked about underground places that go back to the city's beginnings. And so again I wonder about the rat's sense of history: as a rat climbs down through the civic vestiges of man, through the layers of the city that reveal its abundant and varied history, as he makes his way out of the trash and into his nest, does he perceive in some history-powered synapse those ancestor rats, the very first Norway rats, who came on ships from other lands at the time of the American Revolution, who followed these back trails in the past, who fought with the rats that were already here, who colonized and expanded and roamed and, now, in their collective presence in rat history, are the unknowable rat spirit that is part of what makes New York the city that it is? Laugh all you like, but all I knew was that somewhere in this rat hole I would figure out what it was that a year with rats had been trying to tell me.*

Now, late at night, I could feel the ache of history under my feet, the secret in the rat-urine-covered cobblestones whispering to me. Once more I dug into Edens Alley's past, cut down into the history of a hill the way the blade of a saw passes through time as it cuts through the growth rings of a thick, old tree. I headed back to when the hill was still an easily noticeable hill, into the geography that is mostly lost to contemporary humans: the terrain is smoothed yet the hill, still a factor in the rain, still warrants a few extra stairs in a nearby subway exit and is perhaps still noticeable at rat level. I read the old records in the city's archives and saw the hole itself being expanded in 1968 when the Uniformed Sanitationmen's Association put in air-conditioning. I saw the underground vaults installed on the north side of the alley in 1948. I climbed with the rats, in a sense, as time rewound and the worn-down hill rose again, stood up in the ten-block area of office buildings that are filled with large companies but also with small companies too numerous to name. I saw the character

---

* Just a quick note on finding history in holes, on looking down into the past, which is easier than it may sound: Once, in Rome, I went to a church, San Clemente, also known as Saint Clement's Basilica. When you enter the church, you enter a medieval church with eighteenth-century additions—a Baroque basilica. When you go down a level, you see that the upper basilica was built on an early Christian church, with frescoes dating to the ninth century. *That* church was, in turn, built on the site of a mithraeum, a third-century temple for the cult of the god Mithras, which revolved around the life-giving slaying of a bull. (A major feast day of the popular cult of Mithras was December 25, and the cult competed with Christianity for popularity.) You take an ancient set of steps down into the mithraeum, and when you get there—it's a dark, dank, stone-walled, basementlike place—you see in the floor a stream running beneath a metal grate. When you do, you realize that the mithraeum was built on a Republican estate, which was most likely built where it was built because it was alongside a stream, a stream that was probably beautiful at the time, not that I personally have anything against streams piped through basements. My point is this: religions build on religions, cultures on cultures, cities on cities, just the way one rat moves into the previous rat's old rat hole—or hole of any kind, really.

of the neighborhood—and a neighborhood *is* a character, the more you investigate it—as it transformed in reverse from financial services and residences and a giant housing project to a neighborhood of craftsmen and artisans and laborers, in addition to those facilities attending to the more ignominious duties of those trades. I could see back to when Gold Street was the center of the gold industry in New York City, when, in the rat range of Edens and Ryders Alleys, there were men making gold jewelry and making gold leaf. These gold workers, I discovered, came to Gold Street just after the Revolution not because of the hill but because of the swamp. So just as I went to the alley and found myself on a hill, so I looked down from the hill and saw the old swamp out past the housing projects that stand there now, past the Burger King that harvests fast food garbage each night. The last old printers in New York City still call the neighborhood the Swamp even though the old neighborhood isn't much there, and they recall the stench of the tanneries, the leather-making plants that were on old streets—Ferry Street and Jacob Street—that were built over and are now only rat-remembered. The tanners were descendants of people such as Smith Ely, tanner extraordinaire, who came to the Swamp on Gold Street, in 1835, and with other tanners contributed to the malodorous man-made swamp smell, who came to the Swamp because it was an *actual* swamp—a swamp fed by a spring that popped up in September of 1879, when the men building the Manhattan support for the Brooklyn Bridge hit it at a depth of eighteen feet below ground: fifty gallons a minute had to be pumped when the spring was working alone at low tide, and two hundred gallons a minute at high tide. The tanners came, filled the swamp's dreck with more dreck, the rats reaping the discarded results.

But back to the Rat King, who has led me deeper down into the pit—even if I can't capture him in the tremulous beam of my flashlight, in the green light of the night-vision gear, for the Rat King has now taken me down to the time in history when his very first rat ancestors arrived in New York. Now, from my rat vantage, I can see way back in time, to the dawn of the lineage of New York's Rat Kings. I can see, for example, that the land of the rat alley is at the crest of a little valley that runs south to what is today Wall Street. The Leni-Lenape, the first humans known to have inhabited the then *Rattus norvegicus*-free New York, perhaps described it with one definition of the word *Mannahata:* "hilly island." Then again, perhaps they did not: other explanations of the origins of *Mannahata* point out that it could have derived from the word *manahatouh,* which means "place where timber is procured for bows and arrows," or even from *Manahachtanienk,* which means "the island where all became intoxicated," a reference to a time when Henry Hudson landed on the island in 1609 and everyone got really drunk. Certainly, the Dutch saw the hill too, and as high as it was above the swamps and streams, they filled it with wheat, so that when people looked up into it, when the wheat was kissed by sun, the hill appeared golden—in Dutch, *gouden bergh* or Golden Hill. Gold Street was named for Golden Hill. In cities, we are surrounded by hints of the past, such is the richness of nomenclature!

But the existence of an ancient hill is not all that the Rat King showed me. I had followed rats through rent strikes and union movements. Now I was following a Rat King back to the forgotten history of the Golden Hill. I looked at the old maps and read the stories that are no longer read, and I discovered that it was on the Rat King's Golden Hill, on the top of Edens Alley's, that a long-forgotten battle of the American Revolution took place—the very first battle, in fact.

At least sometimes it is called a battle. Other times it is called a riot or just some trouble with a mob. What happened was, British soldiers attacked an unarmed crowd that was just as angry at the soldiers as the soldiers were angry with it. The first man to be attacked was the leader of the colonial masses, Isaac Sears. In the days before the Revolution, Isaac Sears ruled the streets of New York. He is almost completely forgotten, but at the time he was known by the British and the colonials alike as King Sears or just The King. I will tell you just a little bit about him now because Isaac Sears is the hero of Edens Alley, my rat Rosetta stone. See how he precedes all that is ratty in New York and inadvertently summons the very first city rats.

Isaac Sears was born in Cape Cod, Massachusetts, the sixth of nine children, the son of an oysterman. Sears grew up in Connecticut and made his name as a sailor, during the Seven Years' War, a global war in which the colonies mostly fought the French and their Indian allies, and during which New York became rich as a supply port for the English and got used to a certain kind of independence. Sears was a privateer. A privateer was a legalized pirate, who worked freelance for the government, keeping the spoils of plundered enemy ships. He was known for his daring, even among privateers, a survivor of impossible battles, of shipwrecks. Sears was described by a contemporary as a man of "great personal intrepidity; forward in dangerous enterprises and ready at all times to carry out the boldest measures." In 1759 he was shipwrecked on Sable Island, off the coast of Nova Scotia, and saved his crew of nine men. After the war, he settled in New York, marrying Sarah Drake, the daughter of Francis Drake, owner of Drake's Tavern, an alehouse popular among sailors, boatmen, and seaport characters. They had eleven children. Sears invested in ships, which traded with the West Indies and the island of Madeira. When the Seven Years' War ended, a recession hit New York. The British taxed the colonies to make up for revenue lost during the war. As a result, colonial trade with the West Indies came to a halt. People said that the streets in Madeira, once filled with merchants, had grown green with grass.

Isaac Sears became King Sears during the time of the Stamp Act, the first direct tax on the colonies—it required colonists to purchase stamps for all legal papers and was intended to raise money to send troops to America. On the arrival of the stamps, people protested in the streets, so that the governor, Cadwallader Colden, was forced to imprison the protesters in Fort George, the British fort. A crowd then destroyed Colden's expensive carriage and burned him in effigy. On the day before the Stamp Act was to take effect, Judge Robert

Livingston called a meeting at a tavern. Livingston was against the Stamp Act, but as a member of the city's aristocracy, he was also against the disorder represented by a rioting crowd. His plan for the meeting was to convince as many citizens as possible to pledge armed support for the fort. As the meeting opened, the men in attendance paid close attention to Livingston—until Isaac Sears pushed forward, charging that the meeting was an attempt to keep the stamps from the citizens. "We will have them within forty-eight hours!" Sears shouted. The crowd roared. Sears shouted again: "Huzzah, my lads!" Now, Sears turned to Livingston and said, "Your best way, as you may now see, will be to advise Lieutenant Governor Colden to send the stamps from the fort to the inhabitants." In a way, Sears ratted on Livingston. Later, Sears compromised and allowed the governor to turn the stamps over to City Hall; something that people who didn't like him didn't necessarily recognize is that Sears almost always compromised—he understood that issues weren't black-and-white but often gray. From that day on, Sears found that he could exploit his reputation as a mob leader to get what the Liberty Boys wanted without necessarily resorting to force; he would harass people and ridicule them, shout them down in the pubs—he was the people's jerk. Isaac Sears became, in the words of George Bancroft, the nineteenth-century historian, "self-constituted, and for ten years, the recognized head of the people of New York."

Sears reigned as a member of the local chapter of the Sons of Liberty. The Liberty Boys, as they were also known, were a group of workers—sailmakers, printers, shopkeepers, day laborers, a songwriter, fishermen, oystermen, and the tradesmen sometimes called mechanics—who worked in the trades in the city, especially on the docks. (A precursor to the Sons of Liberty in New York was the Sons of Neptune.) In Boston, Liberty Boys were men like Paul Revere and John Hancock and Samuel Adams. They were the revolutionaries just before the Revolutionary War's revolutionaries, the fathers of the Founding Fathers.

Initially, the Liberty Boys didn't want a revolution. They were united in their interest in getting rid of the revenue acts that hurt their businesses. They saw liberty as the freedom to work and make money; the watermark on the stationery of the Sons of Liberty in Albany, for instance, was "Work & Be Rich." They fought for opportunity and the opportunity to work—for many years after the Revolution, the Fourth of July was celebrated as Labor Day. They were protesting for their rights as British citizens, rights they saw not as revolutionary but as standard. To protect their rights, the Liberty Boys encouraged each other's organizations through meetings and correspondence; the Sons of Liberty recommended the establishment of the first colonial postage system, and they arranged the first intercolonial associations. In some cities the Sons of Liberty united with radical farmers in the countryside. (This was not the case in New York, however. "The Sons of Liberty are of opinion that no one is entitled to riot but themselves," wrote a Tory commentator at the time.) The aristocracy shared the Sons of Liberty's opposition to taxes, but to most of the aristocracy in New York the taxes imposed by the British were not as bad as what they called "the leveling principles" and the "democratic notions" of groups like the Liberty Boys. The nonradical rulers of the city called the Liberty Boys all kinds of names. They called them vermin, the mob, *mobile vulgus,* lobsterbacks, Negroes and boys, "flaming patriots without property," "the mixed rabble of Scotch, Irish, and foreign vagabonds," "descendants of convicts," "foulmouthed and sin-flaming sons of discord and faction," "the meanest people," "Children & Negroes," oystermen, and rats. Philip Foner, in *Labor and the American Revolution,* joked that an entire book could be written just using the derogatory names that the upper class called the Liberty Boys and their like as they swarmed through the cities.

The Liberty Boys gathered in taverns, communal places, where they rented pipes and shared cups. Pamphlets, broadsides, newspapers, and handbills were read aloud for the benefit of the illiterate. There were more taverns in New York than in any other colonial city, and the talk in New York taverns was considered especially effusive. "There is no modesty, no attention to one another," John Adams wrote, after he visited New York on his way to Philadelphia. "They talk very loud, very fast, and altogether. If they ask you a question, before you can utter three words of your answer, they will break out upon you again, and talk away." After the British taxed tea, New Yorkers began drinking coffee and taverns were sometimes called coffeehouses but they were still taverns. The Liberty Boys met in Burns Tavern, in Fraunces Tavern, at Drake's, in Montayne's, and in a tavern that they chipped in and bought for themselves, called Hamden Hall. The Liberty Boys also met with women on occasion; they are thought to be the first association in America to have a women's auxiliary, the Daughters of Liberty. The Daughters of Liberty refused to drink tea after the tea tax and boycotted British clothing, saying, "It is better to wear a homespun coat than to lose out liberty." (Once during a Daughters of Liberty demonstration, a man spoke out against American independence, at which point a Daughter of Liberty stripped him of his shirt and, in lieu of tar and feathers, covered him with molasses and the tops of flowers.) A song that the Sons of Liberty sang went like this:

> With the beasts of the wood, we will ramble for food,
> And lodge in wild deserts and caves
> And live poor as Job on the skirts of the globe,
> Before we'll submit to be slaves; brave boys,
> Before we'll submit to be slaves.

They posted handbills around the city that said LIBERTY, PROPERTY, AND NO STAMPS. They erected a Liberty Pole, a flagless flagpole in the Fields, which was also called The Commons and is now City Hall Park—it was the place where New Yorkers gathered and talked and shouted. They said they would "fight up to their knees in blood."

Isaac Sears's great power as the leader of the Sons of Liberty and leader of the revolutionaries of New York was that he could convince so many people to see things his way. Before the Stamp Act and the Sons of Liberty, New Yorkers had no access to government, nothing akin to the public meetings that Bostonians held at Faneuil Hall. Sears used the mob to give people some legitimacy as citizens—for the first time in New Yorkers' history. What the aristocracy saw as riots, the rioters saw as a kind of power; a Loyalist official said, "The mob begins to think and reason. Poor reptiles!" Sears's great tactical success was in helping to foil the Royalist New Yorkers who had promised British officials that New York would desert the revolutionaries' cause; he was the unrelenting rebel presence, always gnawing away at Royalist gains. It has been argued that the British lost the Revolution because they devoted so much time and energy to holding New York. If so, then the leader most responsible for the colonies' ultimate triumph is Edens Alley's forgotten privateer.

As the old revolutionaries became more revolutionary, Sears's Liberty Boys moved accordingly; if Samuel Adams was the first to philosophize about breaking off from England, Isaac Sears was the first to act. With his boats intercepting ships filled with British goods, with his rallies and visits to Tory homes, with his constant verbal and physical harassment of Tories in taverns, Sears is said to have done more to boycott British goods than anyone else in the colonies. In 1765, he sent two Liberty Boys to Connecticut with letters intended to form a military pact between the colonies in the face of possible British aggression—the first move toward concerted physical resistance in the American Revolution. I have never seen a contemporary portrait of him, but I imagine he often had his fist clenched and his mouth open. He was the first in a long line of crowd rulers that subsequently bred Boss Tweed and Tammany Hall and machine-led governments all over America. In 1775, Sears was arrested but the crowds rescued him, carrying him on their shoulders through Wall Street and up Broadway to the Fields. Disgusted that the city did not intercede, a Tory wrote, "Our magistrates have not the spirit of a louse."

After the battles of Lexington and Concord, in 1775, the Sons of Liberty raided the arsenal at City Hall, arming citizens. Sears marched 360 men to the customs house and closed the port. He sent patrols out from his home. He was on a short list of people that the British military called "the most active Leaders and Abettors of the rebellion." His Majesty's ship *Asia* was ordered to attack Sears's home in Beekman Street, due to Sears's success in blocking supplies to that and other British ships. "[F]ire upon the house of that traitor, Sears . . . and beat it down," wrote Vice Admiral Graves. The British put a bounty on his head and tried their best to exterminate him, but at the last minute, just before the British took control of New York, King Sears slipped away.

BUT LET'S NOT FORGET THE rat and the rat alley, because it is there where I stand on a warm evening at the beginning of spring when my Rat King, fed and fought with and triumphant in an overindulged-by-garbage kind of way, waddles down into history—it is in this very spot, I realize at last, that Isaac Sears struck the first blow for liberty, in a skirmish called the Battle of Golden Hill. It was an unglorious blow, an animal-like action, and the first blow in a battle that led directly to the conception of America—as well as to the introduction into New York of the *Rattus norvegicus*. It's an example of the circles of men and the circles of rats closing in on each other, to a point.

The buildup to the Battle of Golden Hill began on January 13, 1770, with a fight at the Liberty Pole, a flagless flagpole that was the lightning rod for both sides' fermenting discontent. The British soldiers hated the Liberty Pole as if it were a living thing; they had already destroyed the pole on several occasions. They had blown up the second pole on March 18, 1767, the anniversary of the repeal of the Stamp Act. The third pole—larger and protected with iron bars and hoops—was destroyed on the night it was erected. This fourth pole stood for three years, but by 1770 relations between the British troops and the Liberty Boys were at a new low. More British troops had arrived in the city, and New Yorkers were being taxed to garrison them. The British soldiers, meanwhile, were given jobs when they were off duty—in the eyes of the citizenry they were taking the colonists' jobs. New Yorkers resented the troops and the troops resented the New Yorkers' resentment of them. The taverns teemed with philosophical arguments and gossip, and people met and talked in the same place where people have always protested in New York City, union leader and ragtag community protester alike—in the Fields around the Liberty Pole, in The Commons. A British military commander in Manhattan wrote to a British commander in Boston, where the situation was similar: "It is now as common here to assemble on all occasions of public concern at the Liberty pole and Coffee House, as for the ancient Romans to repair to the Forum."

On the night of January 13, 1770, forty British soldiers crept out of their barracks, which was only a few yards away from the pole, and attempted to blow up the Liberty Pole: they drilled holes in it and filled the holes with gunpowder. A cordwainer noticed the soldiers drilling. He sprinted into Montayne's tavern, across the street, where a number of Liberty Boys were hanging out. Two Liberty Boys ran out of the tavern to investigate and then scurried back in. The troops lit a fuse but it fizzled. The Liberty Boys came outside again, yelled, "Fire!" to alert the town, then stood and hissed at the soldiers. The soldiers chased the Liberty Boys back into the tavern and then wrecked the place; the soldiers beat up a waiter and chased a customer out the window and threatened Montayne himself. The soldiers tried again two nights later to take down the pole but failed. On January 15, a Liberty Boys broadside, signed by "Brutus," called for a rally at the Liberty Pole the next day. Aside from lamenting the taxes required to house the troops, Brutus argued that New Yorkers were paying poor taxes to "maintain many of their whores and bastards in the work house." "Every man of sense among us knows that the army is not sent here to protect but to enslave us," Brutus wrote. These remarks upset the troops further, and before the Liberty Boys had their rally, the soldiers finally managed to cut the Liberty Pole down. They sawed it up into small pieces and quietly laid them at the door of Montayne's.

It was cold, and snow covered the ground, but when the bell rang on St. George's Chapel the next morning, three thousand people turned out at the spot where the pole had stood. The crowd was mad.

Many of the men present had lost their jobs to the soldiers. The Liberty Boys read a resolution against employing the soldiers, against soldiers roaming the streets at night, against soldiers behaving in "an insulting manner." Violators, the resolutions said, "shall be treated as enemies of the peace of this City." The crowd cheered. "Huzzah!" the people said.

In this primordial moment, at this moment in American history and the history of the city of New York—at this moment that is, in my mind, akin to the moment wherein organic life might have originated in the thermal vents that dot the seafloor—the rivals were face-to-face. A carpenter pointed to a British guardhouse and shouted, "It must come down!" The soldiers standing nearby immediately drew their swords. Bristling with weapons, the soldiers dared the crowd to try to take the house down, and the crowd—growling, roaring—began moving in to do so, until the British officers and city magistrates calmed the two sides. That day, a party of sailors patrolled the streets and docks with clubs, turning out any soldiers they found. On Friday, January 19, the soldiers went out on the street with a broadside of their own. Their broadside argued that the Liberty Boys were the real enemy of the peace of the city; it described the Liberty Boys as murderers, robbers, and traitors "who thought their freedom depended in a piece of wood." The soldiers described themselves as the defenders of English liberties. They said they would not "tamely submit."

King Sears was not happy about the soldiers' broadside. When he ran into a small party of soldiers hanging the paper in the Fly Market—at the intersection of Maiden Lane and Liberty Street, three blocks from Edens Alley—he grabbed the leader of the party by the collar. Sears shouted at the soldier, and according to a report at the time in the *New York Gazette,* he demanded to know what they thought they were doing. He didn't wait for an answer but dragged the soldier to the mayor. Walter Quackenbos, a baker who was a friend of Sears's, grabbed another soldier and followed Sears. A third soldier tried to stop Sears. The soldier drew his bayonet, but Sears had a ram's horn on him and, seeing the sword-tipped rifle aimed at him, threw the ram's horn at the third soldier, hitting the soldier in the head. Amazingly, the soldiers all scattered off, except for the two that Sears and Quackenbos had in hand. Sears and Quackenbos brought the two detained soldiers to the mayor. A crowd quickly gathered outside the mayor's house, and in a few more minutes, twenty British soldiers arrived, their swords and bayonets drawn. A soldier—a colonial soldier who, it was reported, was unemployed because of the British soldiers' presence in the city—went to the door of the mayor's house with a small group of men to turn back the British soldiers. Seeing the soldiers' weapons, people began arming themselves with wooden rungs that they ripped from sleighs. There was shouting. The mayor appeared and ordered the British soldiers back to their barracks. The soldiers obeyed, but as they retreated, the crowd followed them. The soldiers proceeded directly back to their barracks, until they arrived at the foot of Golden Hill, at which point they sprinted up the hill. When the soldiers reached the top, one of them shouted, "Soldiers, draw your bayonets and cut your way through them!" The other soldiers charged, shouting, "Where are your Sons of Liberty now?"

It was a melee, an anarchic moment. Free of civil restraint or control, it was like when you are in an alley full of rats and you stomp, thinking you are in control of the rats, and then the rats freak out and come at you and you end up being freaked out too. As the soldiers and the crowd fought, a second group of soldiers arrived from the barracks. A soldier on the bottom of Golden Hill shouted to the soldiers at the top, saying they should, as one colonial newspaper reported, "cut their way down, and they would meet them half way." The second group of soldiers attacked. The crowd fought the soldiers. A twenty-two-year-old chairmaker's apprentice charging up Golden Hill with a chair leg managed to grab a musket, a belt, a bayonet, and cartridge box, all of which he saved and subsequently used to fight in the Continental Army. In the end, three people were injured, a sailor was beat up, a fisherman had his finger cut off, a water seller was slashed, and Francis Field, a Quaker who had been standing in his doorway, had his face ripped up. The soldiers finally chased the people out into the streets; the crowd scattered, though when other people opened their doors to see what was going on, the soldiers ran after them too. That evening the soldiers attacked two lamplighters, cutting one in the head and pulling the ladder out from underneath the other. The next morning the soldiers came out and attacked a woman on her way to market, and the Liberty Boys broke up a fight between British soldiers and some sailors. Later that afternoon, the soldiers attempted to stop another gathering in the Fields but were beaten back into their barracks by the Liberty Boys. All told, it was two days of vicious scuffles and taunts and armed rioting, a dirty, ratty fight.

The Liberty Boys put up a new pole; it was a mastlike thing, made by seamen, covered with steel sheeting on the bottom and guarded with a fence—British-troop-proof. The pole was carried to the site in a great parade, led by King Sears and his people. The fifth Liberty Pole survived until 1776, when it was cut down by the Loyalist sheriff, who had been whipped at its foot. And two months after the Battle of Golden Hill, after four colonists were killed in Boston in what became known as the Boston Massacre, it was said that an underlying cause of *that* melee was that the British soldiers had been upset by the treatment of their counterparts in New York.

THE ARRIVAL OF *RATTUS NORVEGICUS* in America went unnoted—the opposite of the appearance, for instance, of a rare species of bird. But it seems to me a matter of physics that *Rattus norvegicus* arrived when Sears left. Personally, I believe that it arrived not too long after the fifth Liberty Pole was cut down. In other words, the rats came after Sears had sold his house and moved his family and children, first to Connecticut and then to Boston. It was as if a Rat King vacuum had been filled.

The rats came after Sears evacuated the city he loved in the summer of 1775, along with four-fifths of the population, or about twenty thousand people. In 1776, a third of the city's houses were burned. Then, the city burned again in 1778. Many of the remaining colonists lived in a place nicknamed Canvas Town, a camp of tents and shacks with people living "like herrings in a barrel, most of them very dirty," according to an English correspondent, who added, "If any author had an inclination to write a treatise upon stinks he never could meet with more subject matter than in New York." The occupying troops cut down nearly all the trees on the island, the trees Manhattanites had been so proud to have lining their streets. Meanwhile, the British shipped in German mercenaries; observers noted that the British treated the mercenaries like cattle, prodding and herding them off ships. *Rattus norvegicus* had already invaded Germany by 1776. Consequently, *Rattus norvegicus* invaded America on the German ships of England's invasion force; it was a shadow invasion. The rats couldn't have had less hostile territory. When Britain surrendered New York, the city was all dug up with trenches and garbage was everywhere. It was the perfect habitat for the newly arrived burrow-loving rat—in addition to poverty and injustice, war is good for rats. The black rat or ship rat was already in the city, living in wooden attics and in the holds of American ships, but now the Norway rat arrived and thrived and amassed, eventually rising up from its lowly new immigrant status to rule the city, from a nonhuman-mammal perspective. The newest rat took the throne.

Now that I have seen the original Rat King of Edens Alley, I see Isaac Sears all the time in New York. Since that winter, I see him on the streets, for instance. I see him in the crowds of people standing on subway platforms or in the lunchtime street traffic at busy intersections or even in the trading pits on Wall Street. I see him in the guy who is simultaneously shouting with and leading everyone in chants and laughing louder than anyone else as you leave a baseball stadium. I see him in the woman with a bullhorn on the steps of City Hall, in the people chanting at a protest in Union Square. I even see him in the past: in the bars and saloons and rat fights and riotous mobs in New York; in the riots in Harlem in the sixties, which weren't as bad as the riots in Detroit and in L.A., where I would probably see Sears too, if I looked. I see him with the student demonstrators of the sixties and standing alongside John DeLury, who, like Sears, watched crowds gather and grow and grumble in the very same spot that Sears saw his crowds—City Hall Park, which was once The Commons and before that the Fields and before that a grassy plain bordered by swamps and ponds and in sight of a golden hill. I see Sears in Wisconsin and in Chicago and everywhere there are people, which is everywhere there are rats—people who are sometimes moblike, sometimes the embodiment of justice, sometimes just looking for a good time or something to eat and a decent place to live.

When I look into the rat alley, when I watch the Rat King descend into his dank hole, I see the first New York rat. I see a noble king, a leader of the people, and I see a jerk. I see the beginnings of the great city of New York. I see the great city's forgotten moments and its forgotten people, its bowels. I see rats being killed and then multiplying and then being killed again. I see fear and courage. I see nature and I see human nature and I see great crowds of hungry or drunk or tired or righteousness-inspired men and woman as they rise up and shout, *Huzzah, America! Huzzah!*

THE QUIET CODA TO THE secret of my rat alley is Isaac Sears's fade into obscurity—the story of how his life story skittered off into a hole. I think he may have been forgotten because he once got perhaps a little too self-reliant.

In 1775, Sears rode back into New York with a band of Connecticut horsemen, seizing prisoners along the way, and broke into the print shop of a Tory printer, James Rivington.* Sears and his men sang

---

* Rivington is thought to have been a spy for the Americans—he ratted on his Tory supporters, probably for money. He is said to have delivered to Washington the Royal Navy's signal book in 1781. The spy who was thought to be most valuable to Washington in New York City during the time of the British occupation was Hercules Mulligan. Mulligan was a tailor on Queen Street, the city's most fashionable street, and he was the city's most fashionable clothier, in addition to being a Liberty Boy. When the British occupied the city, British officers came to Mulligan for uniforms, and as they did, he was able to learn about British troop movements, information that he passed on to George Washington. He was thought to be a

"Yankee Doodle Dandy" as they carried off Rivington's type to melt down for colonial bullets—just as the British believed that the colonial rebellion would end if the ringleaders were rounded up, so Sears felt that if Loyalists like Rivington were quieted, then the Loyalist masses would turn to the colonists' cause. It was a huge misstep for Sears. At the time, he was about to be appointed secretary of the American navy. He had been assured of the position. But after breaking up Rivington's presses he was passed over for the job. The young Alexander Hamilton, the next generation of revolutionary, condemned the raid as a defamation of the press. Sears was branded uncontrollable, one of the things he probably was not; George Washington called privateers like Sears "squirrelly," each ship "a free lance." Sears felt betrayed, more so when he asked the Continental Congress for reimbursement for the men on the raid and was turned down. Today, we think of the revolutionaries as being selfless and uninterested in financial gain, but like many colonists fighting the British, the old privateer believed that unrewarded patriotic service was a luxury affordable only to the well-off. "[W]hen a man has done most, he gets least reward," Sears wrote in a letter. He was left to ponder his career-ending moment of unrestrained violence. He

---

traitor by his anti-British neighbors, but he was so important to Washington that when the British left New York and Washington returned to the city, the first thing the general did upon arrival was to have breakfast with Hercules Mulligan and present him with a bag of gold. Everyone in town apparently knew what Washington's gesture meant; it cleared Mulligan of charges of being too friendly with the enemy. According to a short history by the CIA of early American espionage, Mulligan coaxed information out of British generals and got himself out of scrapes with British intelligence by utilizing what the CIA termed "blarney." One reason I mention Mulligan's ratting out of the British is because when my father worked in New York City, his alias was Hercules P. Sullivan; he had business cards printed up with that name and referred to himself as Hercules. I never knew how he chose the name Hercules for himself, but when I ran across Hercules Mulligan in the revolutionary history books, I called up my father and asked if he'd ever heard of him. My father said that one day in the seventies he had come upon a plaque that described Hercules—he recalled seeing it at 160 Water Street, which is near where Queen Street, when it existed, would have been downtown. My father looked at the plaque, read it, and according to his best recollection said, "Holy smokes! That's some handle!" That phrase can be translated roughly as "Wow! What a great name!"

worked with the military through the course of the war, advising on marine matters, devising fortifications for the Hudson Valley, and he returned to New York as soon as the war was over. Like many of his generation, he was forgotten, the younger revolutionaries now leading the way.

After the Revolution, Sears was in debt, like New York. Now that its old British trading partners were gone, the city was looking for new trading partners. Sears arranged one of the first trips to China, believing it would be the key to the new nation's economic success. Sears set sail for China in 1786 on a ship called *Hope*. He came down with a fever and died on the way. He was buried on an island in Canton Harbor.

Somewhere around 1898, a plaque was erected to designate the site of the Battle of Golden Hill. It was on a building at a corner on what I calculate to be the base of the old hill, a block away from the rat alley. That building was demolished and the building that replaced it was called the Golden Hill Building, but the plaque disappeared. In 1918, a reporter investigated the whereabouts of the plaque and discovered that it had been moved a few blocks away—in 1918, in other words, the plaque marked the site of the battle on what was not the site of the battle at all. The man in the building where the plaque stood told the reporter that he had rescued it when the old building was torn down. "I don't know anything about the battle, but I do know it's a handsome bit of bronze and it would have gone to the junk heap if we hadn't looked after it," the man said. Sometime after that, the plaque disappeared entirely and the Golden Hill Building was knocked down and replaced with a building that doesn't mention the battle at all—history papered over, buried, its faint clues fading.

The Liberty Pole still stands, though it is inside a fence alongside City Hall, and if you didn't know it was there you might miss it. It is surrounded by thirteen stones, each said to have come from an original colony. It was replaced after the Revolution and replaced again in 1921, after a ticker-tape parade, then replaced two more times, most recently in 1952, when it was discovered to have rotted out inside.

The Liberty Pole is still surrounded by the half-protective, half-paranoid fencing originally designed by the Liberty Boys. Today, it stands alongside an unidentified stone ruin, a foundation of some kind hidden in the grass. The last time I was there, someone had laid out rat poison right beside it. That makes sense to me. I've seen some rats in there.

## *Chapter 20*

# SPRING

For I have learned
To look on nature, not as in the hour
Of thoughtless youth, but hearing oftentimes
The still, sad music of humanity.
—William Wordsworth, "Lines Written
a Few Miles Above Tintern Abbey"

IT WAS SPRING when I found the dead rat, and I didn't recognize it—though that's not to say it wasn't the rat that I had previously recognized.

I had come by the alley during the day for the very first time. It was a warm, clear spring day—a hopeful day after a fear-filled winter. I was merely walking through the alley, and a little thrown off by the natural light on the cobblestones. Frankly, I didn't expect to see any rats. Then I looked down. At first I thought I saw trash, but then I realized that it was a dead *Rattus norvegicus*. It was on the edge of the alley, roughly where the Chinese garbage berm would have been and would likely be again that evening, though I found it difficult to determine precisely where the garbage had been—everything was so cleaned up on that day that I almost didn't recognize the place. Overall, it was strange to even be there, a little like waking up from a long dream.

The dead rat didn't look as if it had been run over by a garbage truck or attacked by another animal, so I began looking around the alley, investigating, like an exterminator. And then, sure enough, I found it—the rat poison. I looked around some more and saw some more rat poison and put two and two together: an exterminator was working my alley.

The next time I went to the alley at night, there were rats, but fewer rats. The same thing happened again a few days later. Eventually, I went home and called up the exterminator whose name was on the poison bait stations, AA Federal Exterminating. Mike Baglivo answered the phone and politely put me on hold a few times; he sounded pretty busy. I explained—as carefully as I was able—that I had been keeping a rat journal and that he was exterminating in my rat alley. He seemed to understand; he did not seem to have a problem with my experiment. I said I thought he was doing a good job of rodent control in the alley, which, as true as it was, was a little difficult for me to say. I also complimented him on his placement of the bait stations, which were—as I knew, perhaps as well as anyone—in high rat-traffic spots.

"Thank you," Mike said, and launched into a small discourse on rat placement and rat eating habits in general. "You know, the thing with rats is, rats are very lazy creatures, and they'll take food that's right near them rather than go across the street for a steak. It's like that old saying, a bird in the hand is worth two in the bush, if you know what I mean. And the other thing is, they're very edgy. They're always worried, so to speak. They feel they have to keep on the move."

I agreed. Mike went on. "It's a family-run business," Mike said, "which means our guys get out there and do what they have to do. I mean, when you have a small company, it matters more. Not that I wouldn't mind it being a bigger company, don't get me wrong. But when these guys are out there, they know that they've got the account and that if they lose it, we won't make any money. So if one of these guys is on vacation and somebody else does the work and they don't do a good job, then they get kind of sore, if you know what I mean. Our guys go and do what they have to do."

I felt the need to meet with the exterminator who had exterminated the rats in my alley, face-to-face. I asked him if I could stop by.

"I'll be here," he said, and in a little while I was shaking hands with him in his storefront office in the Dyker Heights section of Brooklyn, near New York Bay. The store was full of poisons and traps for sale; the walls were decorated with hunting trophies. I shook hands with Mike, and as soon as I did, he excused himself and was back on the phone. He was in the midst of dispatching four different exterminators throughout the New York area.

"Federal," Mike said, answering. He listened. "Okay, yes, you could see termites for the next two months, but if you see anything next *year,* then call, all righty? Now, it's just a matter of waiting for them to die."

He hung up the phone, but it rang again immediately. "Federal." Mike listened again. "Now, Mrs. S.," he finally said, "what you're doing here is accusing one of my workers, which, if he did do it, he's gonna lose his job. Does that sound reasonable, Mrs. S.? Could you have missed the buzzer?" Mike began nodding. "Okay, we'll send somebody right back. Thank you."

Mike finally said to me, "We do homes, apartments, food processing plants. We do everything." He explained that his father had started the business after working as an airplane mechanic, first in the service and then with Pan Am. His father named the company Federal to sound solid and nationalistic and he added the AA to get the listing up high in the phone book. Mike himself went to pharmaceutical college, but then changed his mind. "When I graduated, it was going down the tubes," he said of pharmacy work. He said he felt that being a pharmacist meant working at a giant chain pharmacy, and he was not interested in that. Instead, he went into the family business, where a knowledge of chemicals helps, given the amount of poison used. Today, he was doing the dispatching, while his father was out on a job.

I asked him about Edens Alley, but there wasn't much to say. He was not the one exterminating the rats there; it was another exterminator who was out in the field at the moment. So we ended up just chatting about rats.

He recalled one of the first rat jobs he ever went out on, in a supermarket in East New York, where the market workers had been shooting rats in the basement with a .22-caliber rifle. Up in the produce section, he noticed a rat sitting in the produce shelf. "It was in the lettuce," Mike said. He thought a woman was about to pick it up, mistaking it for something that was not a live rat; the rat was being misted by a vegetable mister. Mike quickly grabbed the spraying wand of a pesticide sprayer and bashed the rat. "I didn't plan it or anything," he said. "It was just a reactionary move."

I was getting ready to leave—Mike was just too busy. But then Mike was reminded of an aspect of the nature of rats in the city, and as he put down the phone, he said, "You know, I heard there are three layers of sewer lines." He counted them off on his fingers. "There are the ones from the 1800s, the ones from the 1700s, and the ones they don't have the maps for anymore. Once in a while, they use that old line, when they're doing construction or something, and you read in the papers that there are hundreds of rats coming up. Well, those rats that are in the third line, they haven't even *seen* man before. Hold on a second . . ."

One of his technicians radioed in. "Where are you, killer?" Mike said. "What do you figure, fifteen minutes? . . . Okay, you got it." The phone rang again. "You got it, Mrs. Salamo . . . No, it's our fault too, Mrs. Salamo. We'll see you soon." He hung up. "She's a real sweetheart."

As he was talking, an inspector from the state department of environmental protection came into the office to give a surprise inspection of AA Federal's pesticide storage procedures. I shook hands with Mike again and disappeared.

I WAS A LITTLE SAD about the end of the rats in my alley, a little melancholy, and to cheer myself up, I walked down to the alley and looked around. I felt a sort of anticlimactic euphoria, this bittersweet excitement. *I can see the hill!* I thought to myself, as I once again pictured Golden Hill. In the same way, I walked up through City Hall Park and imagined rallies of generations past and generations hence and then saw some bait stations and some people on City Hall's steps and at some point, yes, I could feel myself a part of the crowd. Then I got to thinking about George Ladd, the exterminator who'd sent me downtown looking for rats in the first place, at the beginning of my rat alley experiment, and I walked up to the Lower East Side to see if he was around.

It didn't look as if he was going to be there, but he was—he cracked open his door and greeted me warmly. He seemed to be in good spirits; that big construction project downtown that had been canceled because of the World Trade Center disaster was back on, and he would be doing the rodent control. Also, he was beginning to get involved in pest control politics. He was talking about working with a pest control lobbying group in Washington, D.C.

Just for fun, I asked him if I could see the tape of him on Japanese TV again. It showed him with relics of his great-grandfather's work in Japan; it showed the office cat that had since died; it showed George in his driveway with his wife, as they packed up to go off on a camping trip in the Adirondacks. "He can get upset with things pretty easily," his wife said, "but once you get to know him, he's great." In a part of the program I did not recall, George talked a little bit about why Bonzai de Bug had never become a large pest control firm, why he'd stayed small. "Working for someone's a big thing for me," he said. "I have a problem when it comes to that point when you gotta call manure manure." This seemed to sum up something about being a small-business owner in the city, or even in America, about being a noncorporate cog in a big economic and political and social wheel—about being an exterminator. But as we continued to watch him speak on tape, George said something that cheered me up. "I always say that the money will come if you love what you're doing," he said.

I learned this, at the very least, by sitting in a rat alley: that there is hope in the life of many exterminators. Life is mean and vile in many ways, but exterminators advance toward society's depths and meet life there and see it for what it is, in some cases.

I felt much better after I left George's shop—stopping in on somebody you enjoy talking to will do that. I walked home with rat stories running in my head, and I suppose it will not surprise the reader to hear that, at this point, as I was walking back to my apartment, I began to think more than ever that we are all a little like rats. We come and go. We are beaten down but we come back again. We live in colonies and we strike out on our own, or get forced out or starved out or are eaten up by our competition, by the biggest rats. We thrive in unlikely places, and devour. Our city was not always inhabited, and when we stand in a rat alley, we can see the ancient hills on which our ancestors stood before we infested and devoured the land. We are different and the same; we are touched by the hand of Midas and we are plague-ridden, sons and daughters of Job. We are rats in Congress, rats in a housing complex, rich rats cashing in, poor rats being kicked out. "Rat populations throughout the world are relatively similar, although local conditions and specific differences produce some variation in degree," wrote Dave Davis, and the same can be said for us. We are the rats whose population may boom, whose population may decline, who can survive where no other species could or would want to, in Edens Alley despoiled. With caution, we will flourish; without it we will not; we will starve and die and maybe kill each other, maybe not.

Everyone has heard a rat story, and surely everyone has heard that story that runs through New York and cities everywhere—for I have recounted versions of it in these pages and it is in many ways an emblematic rat story—of the large and scraggly rat that swims through the sewage flow, having gained access through some crack in the underground, through some cobblestone gap, and then rises up in a toilet bowl and infests an apartment building. Who is not disgusted by this? Who is not a little frightened when thinking about its implications toward personal safety and mental hygiene, among other things? And yet who is not impressed that the rat survives—if only to be poisoned or killed with a trap or perhaps a broom or whatever is handy for the person or the pest control operator confronted with this act of sublime magnificence? Does this not give us some hope and concern for our own future as well?

I'm not saying that everyone will agree with me on this point—I'm not even certain it's true, and I don't necessarily think of myself as a rat *all* of the time. But I know that if you look deep down into the darkness, even in a rat hole, there is some life down there, some fecund spark, like it or not.

# AFTERWORD

I HAVE WRITTEN a couple of books prior to *Rats*, but this is my first afterword. I had hoped to write a foreword, but now, since we're here at the end of the book—well, it seems a little late for that. Besides, with a foreword I might scare you off, if I haven't already, just with the title of the book. An afterword seems a little less intrusive, a little more laissez-faire. To me, an afterword says, "If you want to know even *more* about rats, then that's not my problem." Part of me hopes that you have not made it this far, that you finished the book right where I originally planned for you to finish, or even a few chapters before that, and that now you are off leading a less rat-oriented life. The reason being I don't have a lot more to say about rats. I don't mean I couldn't go on and on for pages about them. I just mean that when you write a rat book, when you sit in an alley for a year with night-vision gear, when you type up your notes into a book and then hit the road on a *Rats* book tour that involves you visiting cities all across the country and standing up before perfectly nice people and, night after night, bringing up rats—when you do that you start to worry about what people think of you. As a result, I have a fierce urge to make this afterword and all my subsequent writing about really pretty flowers.

So let me start by throwing out the answers to a few of the questions most frequently asked of rat authors, in no particular order. No, I do not have rats at home. No, my wife does not like rats. Yes, my wife thinks I'm crazy but not *that* crazy. My children think I'm crazy. My parents did not have rats or any affinity for rats that I have ever been made aware of. I have never been attacked by a pack of rats—you give them space, they'll give you space, I have found. I am not with rats right now, typing in fetid squalor, though my desk does need to be cleaned. I don't think wild rats are cute, and, although I have nothing against them, I'm not a huge fan of pet rats either. Yes, rats in alleys fighting over garbage or each other do screech, and they screech really loudly—you can't believe how loudly, the first time you hear them. No, I don't feed wild rats. No, I don't like to lay down and let rats run all over my body. (There are very few things that I like to let run all over my body, as it happens.) And let me make this next answer perfectly clear: I think rats are really, really gross, though through no fault of their own. I think it is our fault, actually. We humans are always looking for a species to despise, especially since we can and do act so despicably ourselves. We shake our heads as rats overpopulate, fight over limited food supplies, and then go to war until the population is killed down, but then we proceed to follow the same battle plan.

Even though I find them disgusting, I *can* relate to rats. In general, rat-book authors usually can. (This has been my experience, though I should point out that I know only one other rat-book author.) The rat-book author is not going to be high on the marquee, and usually the people in charge of the marquee think it's kind of a joke that they have you on it at all. That this book was a bestseller is, I continue to maintain, a kind of ratty mistake. I tell my friends—and I still sort of believe it—that pretty soon there are going to be lots of people showing up at bookstores wanting to return their copy of *Rats* claiming to have mistakenly assumed they were buying *Cats*. Bookstores, by the way, tend to have practical experience with rats. In a California bookstore, I was taken to the very spot where a rat had been, shall we say, discount-tabled by the still-adrenaline-rushed staff; as the booksellers stood there beaming, I felt as if I should have had a citation to give them or something. In Brooklyn, while I was writing this book, my own local bookstore did valiant battle with a *Rattus norvegicus*. I can't remember who actually ended up with the win, but the bookstore, an independent, is still there, an amazing feat of survival these days.

JUST AS ROCK-AND-ROLL BANDS of a certain age like to name their concert tours, so we referred to the rat-book tour at my home as the Rats Over America Tour, and even though I was ostensibly going out on the road to talk to people about what I had learned on the subject of rats, an excellent thing for me was that I got to learn even more about rats in general and about rats in the places where I went. One night, for instance, I was invited to go out into the back alleys of downtown Beverly Hills to look for rats. Rats in California can be, generally speaking, very different from rats in New York. First of all, Los Angeles has two different species of rats: the black rat, or *Rattus rattus,* and the Norway rat, or *Rattus norvegicus*. The Norway rat, as you probably know by now, is the predominant rat in New York City and America. The black rat, which once was the predominant American rat, tends to live in trees and attics, as opposed to Norway rats, which tend to live in basements and sewers; black rats climb along electrical wires and down into Dumpsters. One of the reasons black rats do so well in Los Angeles is that exterminators are better at exterminating Norway rats, which are less sleek and a little easier to trap than black rats; with fewer Norway rats around, black rats prosper. So there I was, after a rat reading late one night, in a suit and tie—I try to dress nice for rat readings to break any rat-author stereotypes that anyone might harbor—and traipsing through Beverly Hills with some very nice people who wanted to see some rats, though didn't *really* want to see rats, if you know what I mean. On that particular evening we did not see any rats, as it happened, but we talked to some workers behind a swank restaurant and we looked through the boxes of a fashion boutique and founds some spots where rats had recently been. It was the same thing in Chicago, except that I got to see rats; in addition, I got to see the back alleys of the blues clubs and lots of art students and beautiful (to me) old buildings that looked

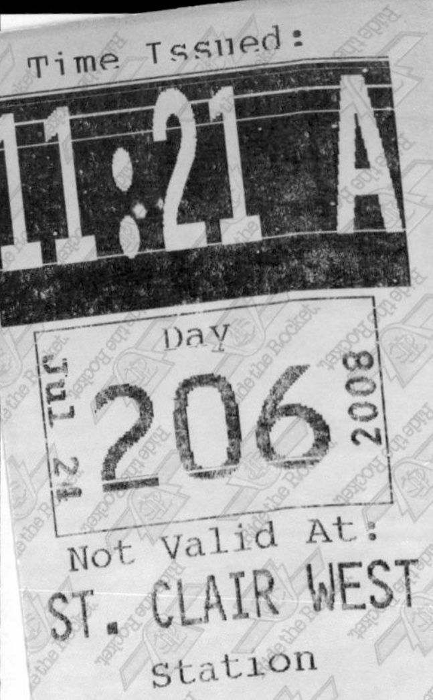


torn down. I also noticed that in addition to
public parks appear to have a bit of a rabbit
pping all over the place, believe it or not. It
t or not.

co, but I also got to know a kind of ratty
ss of searching. I got to find out about a
was happening there. I talked to cops in
at's good (and bad) about the place from a
ve—cops know a lot about rats, as you
ntly surprised to find a very gross alley,
set up shop to sell beautiful handmade
l: rats and tenant organizers and artists are
found together, coexisting semi-peacefully. One night, I retired to a pub that, in addition to serving good locally brewed beers and ales, was at that late hour hosting a small annual convention of letter writers who, I learned, meet regularly to correspond with their legislators—a letter-writers group. I never would have found the place if I hadn't gone ratting. Ratting, for me, then, is not just about rats; it is also about seeing another side of a given city. In other words, I can check into a nice hotel in San Francisco and hit the organic bakery and a few fancy shops, or I can go look at what's going on in the alleys—always with an eye to safety, of course. Checking out rats, I would argue, leads naturally to writing your legislators, or at least to thinking a little more closely about a neighborhood, about where it's going, about where it's been. If you take a map of rat infestation in a city (which is usually a map of where rat bites are reported) and you place it over a map showing the places where somebody ought to be spending more money on social services, more money on repairing the housing stock, or more time just generally caring about the people there, then you will most likely find that they match up pretty closely.

Sometimes, while wandering in big old cities and even in new suburbs, I was called on for advice, and my advice was always the same: get rid of the garbage and the rats will take care of themselves. (I am proud to report that city health officials who attended one rat reading in New York City responded the very next day to a woman who asked me for help with a rat problem in her basement, which sounded to me less like a rat problem than like a full-fledged rat invasion.) In Washington, D.C., people called in to a radio show, sounding sort of defensive to me, to ask if Washington, D.C. rats weren't just as big as New York rats, and I had to tell them, sadly, that while all rats live completely different lives in completely different places, all rats are size- and ferociousness-wise almost exactly the same. But what mostly happened on the Rats Over America Tour was that I got to hear all kinds of excellent rat stories. People sometimes apologize before they tell me a rat story, guessing that I've heard them all. It's true that I have heard an awful lot of them. But I have not heard them all. There may not be one rat per person in America, but there's nearly a rat story for everyone, which is a lot of rat stories. On a talk show a guy called up and talked about a rat in his side alley that he had watched and hunted for months and months, never achieving a final rat bagging, as I recall. There was a woman who, while in bed with a broken leg, befriended the rat that visited her room each evening, even fed him, which sounds to me like something out of a Stephen King story. There were the three men in the financial district of San Francisco who, on their way to work one morning, came upon a rat trapped in a sewer grate. The finely dressed businessmen huddled, knelt down to the struggling rat, and, with their Wall Street white handkerchiefs, successfully freed the stuck creature as onlookers cheered.

My new favorite rat story is not so much a great rat story as a rat story that proves a rat point, and maybe a point about nature. It has to do with the difference between wild rats and pet rats, which is a distinction I thought I made in the book (see chapter 2) but a distinction I had to constantly reiterate while on the road, even to people who were kind enough to have read my book, or to have acted as if they had read my book—as a rat author, I am under no delusions. People often brought pictures of their pet rats to share with me; some people brought their children with their children's pictures of their pet rats, sometimes called fancy rats. And at one reading in Berkeley, California, I thought I was going to have a rat riot on my hands when a small group of people showed up thinking I was against pet rats or something—and, again, I'm not, I swear. It's just that wild rats aren't at all like pet rats—they are *not*, I repeat *not*, cute and cuddly, believe me.

Anyway, I heard my favorite new rat story from a young couple who live in Brooklyn. They showed up at a reading at a cool little bar in lower Manhattan. They didn't say anything during the rats question-and-answer section of my presentation, but afterward the man came over alone, and, pulling me aside, asked if he could speak with me for a moment. I was a little worried about what he was going to ask me—I'm no good, for instance, at relationship advice. But he eventually explained that his girlfriend worked for an animal-welfare organization. She had adopted a rat that had been rescued from the World Trade Center and handed over to her animal-welfare group; the caged rat had survived the towers' destruction. The man recounted to me how, after a while, his girlfriend took the rat home. It developed cancer and died a short while later, but they had both grown accustomed to the rat and were saddened by the empty cage. One evening, while walking home through the streets of the Bedford-Stuyvesant section of Brooklyn—a neighborhood that has seen its ups and downs and is lately beginning to see some ups again—the man spotted a small rat on the sidewalk, a juvenile. He decided to bring it home to the empty cage, as a gift to his girlfriend. He caught it with his hands, noting immediately that it was more aggressive than their previous rat. At home, the rat grew quickly, but while the couple had assumed it might mellow, it did not. Indeed, the opposite happened. The man said that when he placed food in the cage, he made certain to quickly jerk out his hand; he likened feeding the rat in the cage to feeding a piranha. As the man recounted this rat story to me, his girlfriend finally approached, and soon they were both describing their fear of their new "pet" rat. I say *fear* because their eyes beseeched me, hoping I would understand, and, frankly, I did, because I've been there, sort of—I mean, I've been with wild rats. They told me that they wanted to release the rat in a park—coincidentally, a park in the borough of Queens that I knew from my childhood. But they were afraid the rat might jump out of the cage and immediately turn and attack them. I told them to be careful; I told them to play it safe. I suggested that one release option might be to open the cage and run like hell. I also told them that I was glad to hear a story that proved once and for all the difference between the wild and pet versions of *Rattus norvegicus*—proved it for me, at least. In my own mind, I equate the difference between wild *Rattus norvegicus* and fancy *Rattus norvegicus* to the difference between American *Homo sapiens* and European *Homo sapiens*—same species, completely different upbringing.

THAT'S ABOUT ALL I have to say afterward. I suppose I should add that I still drop by my rat alley once in a while, for old times' sake. Sometimes there are rats there, sometimes there aren't. (These days, they are easier to see *in* the garbage: since I began ratting, see-through garbage bags have become more prevalent, and a boon to the rat observer.) Mostly, as far as it goes afterward for me and *Rats*, I'm still kind of amazed. *After* you write a book and it is published and months later you pick it up one more time and thumb through it and close it and just look at its excellent rat cover, you think again about all the nice people that you got to meet and talk to and work with, merely by asking some questions about rats. There are the exterminators and pest-control people working with city governments who don't get a lot of face time on TV but are ready to give you all the time you want when it comes to spreading the rat word. They invite you to rat conferences, where they recall exciting rat situations and talk openly about unsolvable rat issues. Afterward, even my neighbors invite me over to look and see if they have rats, which they more often than not do, and I'm not just saying that because since I wrote *Rats* these same neighbors, when they see me across the street or on a train or across a crowded room, yell, "Hey, Rat Guy!" I'm saying it because they really do have rats. Remember: Rats are everywhere. Don't think they're not near you.

So I want to say thank you to all those people here and say thank you as well to a guy named Dan Millner, who adapted—which in this case means *cleaned up the language of*—a great rat song, "McNally's Row of Flats." I heard him sing it once with Bob Conroy at the South Street Seaport, just down from my rat alley, and since then I have grown to enjoy singing it whenever I am out talking about rats. It was written by Edward "Ned" Harrington, a comedic actor and minstrel singer, in 1882, the music composed by his father-in-law. It's about living in a tenement where more languages are spoken than in the ancient city of Babylon, where rent might be collected by the taking of the bedding and the slats, where things are flea-infested and laugh-infected and crummy but still to some extent good. And it's about what I see *Rats* as being about, which is a bunch of different kinds of people all swarming together in the places where they have historically swarmed together and having either a good time or a bad time but definitely having a time of it. Here is the chorus. If you happen to know the tune then you can sing it after this afterword. Otherwise, just shout it out:

*Ireland and Italy, Jerusalem and Germany,*
*Chinese, Africans, a paradise for rats.*
*Jumbled up together, in the snow and rainy weather*
*They constitute the tenants of McNally's row of flats.*

Now, go and have a drink or relax or something, because the book you just read that was all about rats is thankfully over.

# NOTES

## CHAPTER 1: NATURE

I learned about John James Audubon and his days in New York from two biographies, *John James Audubon* by Alexander Adams and *Audubon* by Alice Ford. According to Ford, when Audubon finished his second book, *Viviparous Quadrapeds of North America*, he took it to Congress, hoping the government would purchase it, but, as Ford writes, "[t]he more innocent members mistook the squirrels for rats . . ." I also read a collection of essays about Audubon, *The Bicentennial of John James Audubon*, published in 1985, and according to the essay entitled "The Dream" by Alton A. Lindsey, two hundred of Audubon's paintings were damaged by a family of Norway rats while they were being stored. When Audubon died, his wife sold his paintings, which were again being eaten by rats, just so that she would have money to live on. At first, she couldn't sell any of them. Then she sat down with officials of the New-York Historical Society and pointed out the importance each painting, one at a time; the society finally bought them. Audubon's grave is in Trinity Cemetery in the Washington Heights section of Manhattan, which is on land that was once his. The cemetery is near the former site of Audubon's house, at about 155th Street and the Hudson River, and the site is one of those many places in the world where you can see the past without trying hard, even though there is no trace of the house, which, like Audubon, fell down and then was covered over by apartment buildings and then the edge of a viaduct. In 1923, according to newspaper reports, twelve families were living in squalor in the three-story house that had been Audubon's home—there was trash on the verandas, and where Audubon once had caged deer and other wild animals were pigs. The studio—according to a *New York Times* report on April 23, 1923, entitled HOUSE AUDUBON LIVED IN FAST FALLING INTO RUINS—was a home in itself: "Today, the room is the kitchen, bedroom and parlor of an aged old woman, who looks blankly at you, when you ask if you may see Audubon's studio." To get to the onetime site of Audubon's house, I took the Number 1 subway train to Washington Heights. (Malcolm X was killed near Audubon's old house at the Audubon Ballroom.) To see the rat holes, I used Nikon binoculars. To see in the dark, on my early wannabe-Audubon trips, I used night-vision gear manufactured by Night Owl Optics, a company based in Manhattan.

## CHAPTER 2: THE CITY RAT

Naturally, I read every rat book I could get my hands on—and for a while I couldn't get out of a library without checking all their various wildlife encyclopedias for one more entry on rats—and they mostly say the same things. But chief among the books I referred to over and over while watching and reporting on rats was *The Brown Rat* by Graham Twigg, an exacting work written by a scientist for the benefit of the lay reader that is gloriously serene in its nonhysterical description of rats. The other book that I referred to most frequently was *Rodent Control: A Practical Guide for Pest Management Professionals,* Robert Corrigan's rodent control manual. This book is written specifically for people working in the field of pest control; it speaks of rat problems pro forma—and as such it is full of clear-cut rat habitat nuance. A good book about rats for the lay reader is *More Common Than Man: A Social History of Rats and Men* by Robert Hendrickson; I referred to it especially for insight into the role of rats in literature and music and rat lore in general. For instance, it describes rat tortures referred to by Freud in his famous analysis of the so-called Rat Man—for my own research, I didn't want to go there. *The Rat: A Perverse Miscellany* is a collection of rat-image and rat-related prose and poetry; it includes a picture of a Rat King, which is disgusting, on page 48. The best book ever written about catching rats is *Tales of a Rat-Hunting Man* by D. Brian Plummer, who has been called "the most famous rat catcher in Britain," but I also found it helpful in understanding rat habits, in addition to the habits of rat catchers. Plummer writes, "Rat hunters are usually regarded as some kind of lunatic by the public at large, and, on reflection, the public at large is right."

As far as scientific articles written about rats go, I am indebted to William B. Jackson, a longtime professor at Bowling Green State University, who sent me numerous scientific articles, written by him and by others. Of articles written by Jackson himself, especially helpful were "Rodent Behavior," which was published in *Cereal Food World,* in 1980; "Food Habits of Baltimore, Maryland, Cats in Relation to Rat Populations," published in *Journal of Mammology,* in 1951; and "Norway Rat and Allies," published in *Exotic Species,* in 1982. Jackson's article "Rats—Friends or Foes?" published in *Pest Control,* in 1980, while less academic, is a good place to read about the American public's disdain for rats; Jackson cites, for example, an episode of *The Tonight Show* in which Johnny Carson displayed a clear plastic mousetrap with its own gas chamber. "When he attempted to demonstrate its utilitarian function," Jackson wrote, "the studio audience booed. Then he asked: 'How about rats?' The enthusiastic audience response was, 'Yeah! Yeah!' "

I spoke with James Childs early on in my alley studies and he pointed me to dozens of helpful articles, including "Seasonal and Habitat Differences in Growth Rates of Wild *Rattus Norvegicus,*" which he cowrote with Gregory Glass and George Korch—the article refers to rats as a "cosmopolitan species," a great phrase. Childs, who studies reservoirs of animal diseases at the Centers for Disease Control, also once experimented with a method of calculating rat populations via rat bite data; he worked in New York City. In talking to Gregory Glass, I learned that another way to think about the diseases that rats carry is to think about the diseases that they carry that we don't even know to check for—rats as vectors of the unknown. Both Childs and Glass have spent time in the modern alleys of Baltimore trapping rats, using Tomahawk traps and peanut butter. Just in speaking briefly with Childs, I know he has great rat-trapping stories.

Various rat facts also came from columns in *Pest Control Technology* and its special sections on rats; I read Robert Corrigan's theory of why rats gnaw in "Rodents' Annoying Gnawing Habit," an article in *Pest Control Technology* in 1997. I found the figure of fifteen thousand rats resulting from the one mating rat pair in a *National Geographic* article entitled "The Rat: Lapdog of the Devil" by Thomas Y. Canby, published in July 1977. (A rodent expert who was present for a photo shoot for *National Geographic* described how a photographer photographing rats under the bed of an Italian woman in Italy had a problem when the husband of the woman returned to the apartment and didn't believe that they were just photographing rats.) In Hans Zinsser's book *Rats, Lice, and History,* a classic 1935 investigation of the effects of disease in history, Zinsser gives as an example of the human's tremendous fertility rate the case of Samuel Wesley, a seventeenth-century Englishman. Wesley had fourteen children by Sukey, his first wife. He left her, reconciled with her, and sired five more children, the oldest of the five being John Wesley, the founder of Methodism. For information on the rat eradication project on New Zealand's Campbell Island, I referred to news reports from the BBC. Information about so-called fancy rats came from pet rat associations in the U.S. and Britain, including the American Fancy Rat and Mouse Association. A good Web site to find out about pet rats, if you are interested in that kind of thing, is rodentfancy.com. The first U.S. fancy rat club appeared in 1978. The recent appearance of monkeypox in the U.S. may have been initiated by a three-pound pet giant Gambian rat, which infected a prairie dog, a species also known to carry bubonic plague, while they were together in a pet shop.

In researching the story of the settlement of *Rattus norvegicus* in America, I read "The Introduction and Spread of House Rats in the United States," an article written by James Silver of the U.S. Biological Survey that appeared in February 1927 issue of the *Journal of Mammalogy,* a scientific journal from that period that is enjoyable to read because it combines personal observations ("One day I was in Mosquito Gulch, and at a miner's cabin, at 11,500 feet, was a bison skull," a report entitled "Altitude Limit of Bison" says) with mammalogist-oriented news reporting; article titles include "A Possible Albino Armadillo," "How Do Squirrels Find Buried Nuts," and "An Interesting Deer From Szechwan." I also read "Entrance and Migration of the Norway Rat into Montana," which was published in the *Journal of Mammology*, in 1947, by which time the spelling of *mammalogy* had changed to *mammology;* the article was written by Clarence Archer Tyrone Jr., who was at the Montana Agricultural Experiment Station in Bozeman. (A 1956 article in the same journal discusses Norway rats in Nome, Alaska; it indicates that Nome-based rats' legs are often frostbitten and that they seem to have more than the usual amount of rat hair, but that they also have fewer parasites than rats in the lower forty-eight states.) I learned that the Canadian province of Alberta considers itself rat-free via e-mail from officials in the Alberta agricultural ministry and, initially, from a publication of the Alberta department of agriculture, food, and rural development, entitled "The History of Rat Control in Alberta," which includes this statement: "Thus, the people of Alberta are extremely fortunate not to have rats." Alberta did have rats in its border areas for a brief period, and at that time, one Alberta mayor refused to believe it. He stated that he would eat any rats found in his town and only changed his mind when he was presented with a bushel full of *Rattus norvegicus*.

## CHAPTER 3: WHERE I WENT TO SEE RATS AND WHO SENT ME THERE

Dave Davis's team at Johns Hopkins University was known as the Rodent Ecology Project, and they worked out of the Department of Parasitology. "Studies on the Home Range in the Brown Rat"—probably my favorite Davis paper, due to all the drawings of rat trails in back alleys—was published in August 1948 in the *Journal of Mammalogy* and coauthored by John T. Emlen and Allen W. Stokes, two apparently key members of the rodent ecology team. This paper contains a statement that caused me to seek out an alley to study rats, as opposed to just anywhere: "Although rats cross alleys, they seldom cross streets." Dave Davis wrote many other rat papers with John Emlen and Allen Stokes, including "Methods for Estimating Populations of Brown Rats in Urban Habitats," which was published in *Ecology* in 1949. A work that Davis wrote alone was "The Characteristics of Global Rat Populations," which appeared in the *American Journal of Public Health,* in February 1951. Davis himself debunked the "one rat per person" myth in "The Rat Population of Baltimore, 1949" and "The Rat Population of New York, 1949," both published in 1950 in the *American Journal of Hygiene.* (Jackson revisited Davis's revisiting of the one-for-one statistic in a 1992 article published in *Pest Management,* "How Many Rats Are There?") In these articles, Davis also showed that rats tended to be in areas where apartments rented for less. That is still the case today. In fact, if you look at a map of mouse and rat infestation in New York City, the highest areas of infestation roughly match up with the areas of greatest prevalence of poverty, not to mention disease and illegal drug use and all kinds of problems—rats are sometimes an indicator species for people who are having a tough time.

In reading about Davis's work at Johns Hopkins, I also read about the history of Johns Hopkins, the first independent, degree-granting institution for research and training in public health—and probably the first place to use rats extensively in public health experiments; Johns Hopkins School of Public Health—now renamed the Bloomberg School of Public Health, for the current mayor of New York City, Michael Bloomberg, who is a major donor to the school—turns out to have been a natural habitat for rat research. Some of the first experiments on rats made at Johns Hopkins were done as a part of nutritional studies in the early 1900s, by Elmer V. McCollum. McCollum was the first celebrity nutritionist; he was referred to as Dr. Vitamin by *Time.* He made his first experiments with rats prior to arriving at Johns Hopkins, when he was at the University of Wisconsin. In Wisconsin, he kept his rat work secret because the Wisconsin state legislature would not support public expenditures on the room and board of rats, a pest to the Wisconsin farmer. Meanwhile, working with rats in a laboratory was considered crazy; McCollum had initially attempted to experiment on wild rats, but, in his words, "they proved too savage to maintain in the laboratory." According to *Disease and Discovery: A History of the Johns*

*Hopkins School for Hygiene and Public Health, 1916–1939* by Elizabeth Fee, McCollum proposed putting vitamins back in bleached flour, extolled the virtues of fruits and vegetables, and was an adviser to Clarence Birdseye, the frozen-food magnate. In a way, McCollum's work with rats helped spawn the modern women's magazine, or at least the modern women's magazine cover, which subsequently created the modern men's magazine cover; his research into early dietary supplements spawned such self-help-like articles as IS YOUR BABY RUNNING THE RISK OF SCURVY? and ARE THERE SUCH THINGS AS NERVE FOODS? and MY HUSBAND SAYS I'M HARD TO LIVE WITH.

In studying the work of Dave Davis, I also spoke to several of Davis's former colleagues and students, including William Jackson, who lives in Bowling Green, after having retired from Bowling Green State; P. Quentin Tomich, a biologist in Hawaii; and Jan O. Murie, who is at the University of Alberta. Jackson mailed me a copy of Davis's unpublished paper "Agricultural Expansion and Intellectual Ferment," which included a forward by Davis's daughters describing him up early watching starlings. Bruce Colvin, a rat expert who is probably best known by the lay public as the man in charge of rodent control during the construction of Boston's underground highway, the so-called Big Dig, explained to me that toward the end of his career Davis became frustrated with civic rodent control efforts; Davis felt that true rat control required a political will that politicians were not always able to summon, or command. As a private rodent consultant, Colvin has, in a sense, taken up the baton from Davis in the area of governmental rodent control—Colvin speaks about rats to government officials, in the U.S. and abroad. (In advising the city of Washington, D.C., for example, he suggested that the rodent control program be moved from the department of sanitation to the department of health.) I attended a "Rat Summit" at Columbia University sponsored by New York City councilman Bill Perkins, who has worked hard on rat issues in New York—many people believe that many rat problems could be solved with the reintroduction of metal garbage cans, for instance, as opposed to easily rat-breached plastic bags, though landlords oppose this due to the weight of the garbage cans. Colvin used a slide to show where rats live in urban areas; in one slide a woman was eating lunch next to a series of rat burrows. "When you walk around the city, you will see it differently after you see this presentation," Colvin said. He also said, "New York is a great place. It's a big city. You have a lot of work to do but you can win. Thank you very much."

## CHAPTER 4: EDENS ALLEY

Around the time of my first trips to Edens Alley, I read about George Trumbull Ladd, the ancestor of George Ladd; I read of his career in the *Encyclopaedia Britannica* and in papers given to me by George Ladd. (In an essay I read, George Trumball Ladd described the human being as "an organism with a mind purposefully solving problems and adapting the self to its environment.") The descriptions of the old Theatre Alley and information about Melville and Whitman and Thoreau walking through Theatre Alley and its environs comes from an amazing pamphlet called

"Four Literary-Historical Walks" by Elizabeth Kray and published by the Academy of American Poets in 1982. It includes maps and descriptions of particular addresses as they would have appeared in each author's time. Kray notes that Melville and Whitman probably stood at the same spot on the Battery to admire the view and that Poe and Whitman both had their skulls examined at Fowler's Phrenological Cabinet, "[b]ut otherwise the three writers had little or nothing to do with each other."

The history of Edens and Ryders Alleys are not, as I note, well documented, and while realizing that, I read the street-history work of Charles Hemstreet, who wrote *Nooks and Corners of Old New York* in 1899 and *When Old New York Was Young* in 1902. Both works have lots of things to say about streets besides Edens Alley, such as Saint John's Lane, a six-hundred-foot-long old street in the TriBeCa neighborhood that still exists and was named for the adjacent St. John's Church; of Saint John's Lane, Hemstreet says, "St. John's Lane is so completely forgotten that in years its name has not even crept into the police records." I found some small bits of information about Edens Alley in *Stokes' Iconography*, an incredible six-volume collection of news reports, maps, and minutes from city government meetings, not to mention photos and all other kinds of historical notes, that were combined into several volumes by I. N. Phelps Stokes, who is described by *The Encyclopedia of New York* as an architect, historian, philanthropist, and housing reformer. The will of Medcef Eden is in the abstracts of wills found in *Collections of the New-York Historical Society for the Year 1906* (vol. 15, 1796–1800), page 128. I found information on Eden himself in "Romance of the Historic Eden Farm; Owned by Astor Family Since 1803: In his Will Medcef Eden Gives Picturesque Account of its Rural Charms in 1798—How Henry Astor Obtained His Portion When Title Was Finally Cleared After Long Litigation," an article in the *Times*, published on February 29, 1920, and in "New York's New Up-town Centre," from the *Times* of September 21, 1902. John Rider is mentioned in *History of New York State: 1523–1927*, volume 5, *The English Period*, by Dr. James Sullivan, a publication of the Historical Society of the Courts of the State of New York.

I also consulted *Old Streets, Roads, Lanes, Piers and Wharves of New York, Showing the Former and Present Names Together with a List of Alterations of Streets, Either by Extending, Widening, Narrowing, or Closing* by John J. Post, which was compiled in 1882. I also read *The Historical Atlas of New York City*, by Eric Homberger, and I looked at scores of maps of the Edens Alley neighborhood in the Map Division of the New York Public Library, where the librarians were helpful in leading me through New York map history. Meanwhile, a fun Web site for investigating old alleys is www.forgotten-ny.com; the site has a nice photograph of Ryders Alley while the apartment building at the very end was undergoing construction. (Also, see that Web site's pages on streets you can visit with their original paving stones.) In researching my rat alley's environs, I am indebted to the work of Charles Lawesson; Lawesson looked at every street in Manhattan and compared its place on every one of the city's old maps. He then collected the history of each street in several small but thick, unpublished, loose-leaf binders that he subsequently donated to the New York Public Library's map division—an encyclopedic task. Although I have never met Lawesson, I have a hunch that he knows the streets of Manhattan better than anyone else in the city.

Information on the history of the tree of heaven came from *The Urban Naturalist* by Stephen Garber. I learned about Jens Jensen, the early proponent of native plants, in an article entitled "The Mania for Native Plants in Nazi Germany" by Joachim Wolschke-Bulmahn in *Concrete Jungle: A Pop Media Investigation of Death and Survival in Urban Ecosystems*, edited by Mark Dion and Alexis Rockman, and also in an article entitled "Natives Revival—Is Native-Plant Gardening Linked to Fascism?" by Janet Marinelli in *Plants & Gardens News,* summer 2000.

## CHAPTER 5: BRUTE NEIGHBORS

The rat infestation in the Flatlands was reported in the *Times* on August 21, 1969, and the Brooklyn trolley infestation was reported in the *Times* on July 2, 1893, in a report entitled "Brooklyn's Plague of Rats: The Introduction of Trolley Cars Drives Them Into Dwelling Houses." In 1949, Mayor O'Dwyer said, "Something should be done," according to the *Times* of April 12, and according to the *Times* of June 28, 1950, O'Dwyer appointed Colonel William A. Hardenbergh as the sanitary engineer in charge of New York's rats. (Not too long after that, O'Dwyer was driven to Mexico instead of facing charges of corruption.) The Baruch Housing infestation was reported in all the city's newspapers in the summer of 2000. The Associated Press reported that antirat demonstrators at City Hall that summer chanted, "One rat, two rats, three rats, four . . ." The Rat Summit—the same one at which Bruce Colvin spoke—was held on November 29, 2000, at Columbia University; I sat in back, and a guy on his cell phone in front of me kept taking calls and saying, "Guess where I am? I'm at the Rat Summit!"

The Rikers Island rat story was reported on many occasions in the *Times:* "Rikers Island Rats Trap and Kill Dog" ran on August 29, 1915; "Rikers Island Rats Face Gas War Today" on September 10, 1930; "Shroeder to Direct Gas War on Rats Swimming from Rikers Island to Roslyn" on September 31, 1930; "City to Forbid Rat Shoot" on May 26, 1931; "2,000 Rats Poisoned in Rikers Island War" on April 12, 1933. The 1933 article began like this: "All was quiet yesterday along the waterfront of Rikers Island." On September 13, 1915, a letter to the editor discussed the use of snakes and bacteria. And I read about the closing of the dump on Rikers Island in "Garbage! The History and Politics of Trash in New York City," a pamphlet by Elizabeth Fee and Steven H. Corey prepared for an exhibit of the same name held in 1994 at the New York Public Library. "Garbage!" reports that after a failed attempt to deodorize the entire island with treated seawater, Rikers Island's rat problem subsided when the city closed the dump in 1933. The city closed it so that the ashes from the burning garbage piles would not blow onto people attending the World's Fair, in the Corona section of Queens, which, before the World's Fair, had also been a garbage dump. I read about the general history of Rikers Island in the *Encyclopedia of New York* and the *WPA Guide to New York.*

The death-oriented rat stories came from the *Times* as well. The story of the Irishman who committed suicide was published on September 25, 1886, and the story of the man who ate rat poison ran on December 4, 1899. The florist harpooned the policeman on July 31, 1897, the story appearing in the next day's *Times*. The story of the exterminator named Walden accidentally fumigating an old woman to death appeared in the *Times* on October 1, 1913. The Park Avenue rats story was covered extensively by the *Times* in 1969—they even ran photos of the rat burrows, the scene was so appalling. To its great credit, the *Times* noted that federal aid to rat-infested impoverished neighborhoods did not move as quickly as the city's exterminators moved against the Park Avenue rats. "The speed with which city officials have responded to news that a colony of rats infests a traffic island in a posh section of Park Avenue contrasts with the callous attitude exhibited not long ago by some members of Congress opposed to Federal aid for extermination programs," an editorial said. A Southern Democrat, the editorial noted, said, "Mr. Speaker, I think the rat smart thing for us to do is to vote down this bill rat now." The editorial went on, " 'The rat smart thing,' and the only humane thing, to do is to escalate the battle against rats. Park Avenue's new terror has been a frightening way of life for too much of the city for too long."

The *Daily News* story headlined "U.S. Experts Wander at Red Show & Wonder at Nothing" ran on June 30, 1959, and the death of the baby who was bitten by rats was covered by all the papers in 1959. The series of articles about the *Daily News*'s own rat extermination campaign ran over the summer of 1960. I first read about the *News*'s so-called rat patrol on March 19, 2001, on the *Daily News*'s daily history page; that article also mentioned that between January 1959 and June 1960 the number of rat bites in New York was twice as many as were reported in America's next ten largest cities combined. The letter signed "Precautious" describing rats in Morningside Park was published in the *Times,* and the doorman's description of "big five-pounders" was also in the *Times,* on March 24, 2002. The night watchman's description ran in the *Times* on February 5, 1889, and the man who said, "*This* is what's happening," and presented a rat to Mayor Lindsay was quoted by John Kinfer in a rat-related dispatch that ran in the *Times* on August 6, 1967.

The woman who was attacked by rats in Edens Alley was big news in the city at the time, naturally—the incident was described in the *Times*, the *News,* and the *New York Post* on the day after the attack occurred; the *Times* headline was "Scores of Rats Are Found at Site Where They Attacked a Woman." Pete Hamill, then a *Post* columnist, wrote a column about the attack—"In the City's Rat Race, Theirs Is the Marathon Event"—and it described the work of the laborers in the rat-infested pit, including Larry Adams, now the city's senior exterminator—it seems as if everyone writing anything about rats has for years written about Larry Adams. When I interviewed Adams, he told me that when rats started running in the hole, Hamill did not run away, like all the other reporters. (Articles in the papers during the same week mentioned that the human population of the city had declined by 4 percent to less than 7.2 million.) I read about the explosion of the bar on Ann Street in the papers on December 12, 1970; the *Times* headline: "60 Injured in Blast That Shatters Bar off City Hall Park." I talked to Randy Dupree, the city official in charge of rodent control at the time of the Ann Street infestation. I first read about Christy Rupp in a May 12, 1979, *Post* article entitled "City Rodent Corps Blitzes Dirty Rats of Ann Street"; Rupp is an environmental sculptor, who still sells now, on her Web site, some of the same rat images that she had been putting up near Theatre Alley in 1979. Her Web site notes, "I started pasting these up during the 3-week garbage strike in May of 1979. Never intending to defend rats, I wanted to point out how we had created a habitat for them, and they would naturally occupy it. The city has it's own ecosystem with a delicate balance. Rats were very visible in those days where I lived in the Wall St. area. Especially around dusk when the human traffic would abruptly taper off leaving all the day's harvest for the first rats to discover. During this time I studied rat behavior and found them to be similar to people in many ways, not least of which was the ability to work together as a community, making them possibly better suited to living in NYC at times." A rat-related news item that I did not mention was the radical magazine *RAT* of the late sixties and early seventies. Francis X. Clines, a *Times* reporter, described *RAT* in a December 5, 1970, article as an underground newspaper that closely followed revolutionary movements.

## CHAPTER 6: SUMMER

I learned about cockroaches from *The Urban Naturalist*, and *The Compleat Cockroach* by David Gordon.

## CHAPTER 7: UNREPRESENTED MAN

The quote from Emerson at the opening of the chapter concerning Jesse Gray is from his essay entitled "Representative Men." Some of the many articles I read in the *Times* while researching Jesse Gray's life include "Harlem Slum Fighter," December 31, 1963; "Rent Strike Due to Double in Size," February 1, 1964; "Rent-Strike Chief and 10 Arrested," February 8, 1964 (by Homer Bigart); "Rent Strike March on Police Planned," March 22, 1964; "Forged Petitions Are Laid to Gray," August 21, 1965; "Jesse Gray Arrested in Rent Dispute," June 24, 1966; "Warrant Out for Jesse Gray," October 14, 1966; "Lindsay Says Gray Is Not an Extremist but Assails His Bias," October 8, 1969; "Jesse Gray's Son Charged with Possession of Cocaine," February 19, 1970; "Gray Withdraws His Mayoral Bid," April 14, 1973; "Jesse Gray Is Evicted for Not Paying Rent," June 23, 1973; "Some Protesters Have 'Burned Out,' Others Say They're Regrouping," September 30, 1977. The headline for the brief, 1982 obituary of Jesse Gray read, "Jesse Gray, 64, Leader of Harlem Rent Strikes." Other *Times* headlines from the time of the rent strike include "Harlem Sit-in at City Hall Wins Promise of Heat for Tenements," "Rat Bites Son of Harlem Rent Striker," "200 Rubber Rats Sent to Governor," and "Governor Sends 'Rat' to Wagner: 'Reroutes' Symbol of Slum Protest to City Hall." The article about Jesse Gray's speech on the day before the riot is entitled " 'Guerrilla War' Urged in Harlem: Rent Strike Chief Calls for '100 Revolutionaries,' " by Junius Griffin, July 20, 1964. This was followed by "Gray Denies Role in Inciting Riots: But Court Continues Ban on Harlem Leader's Group," as reported by Peter Kihss on July 30, 1964, in which Gray disputed the *Times* article by Junius Griffin, saying the tape recording of Gray had been doctored. In "McKissick Criticizes Moderates and Voices Defense of Rioters," in the *Times* on August 1, 1967, Floyd McKissick, the executive director of the Congress of Racial Equality and the onetime leader of the Freedom Riders—the groups of blacks and whites who rode through the South on buses fighting discrimination in public places—held a press conference in Harlem after riots and urged a defeat of a federal antiriot bill, support for a ten-billion-dollar increase in funds for the Office of Economic Opportunity, support of inner-city black-owned businesses and a black university that concentrated on "black culture and black history," and resumption of the urban rat control legislation in Congress. "History will likely record the explosions of this summer as the beginning of the black revolution. The criminal connotation of the term *riots* will be erased. They will be recognized for what they are—rebellions against oppression and exploitation," McKissick said. I found an article compiled from various wire services in the August 8, 1967, issue of the *Milwaukee Journal* about the time Jesse Gray was arrested in Congress for carrying in rats. (An editorial in the *Journal* called the protest "trouble for trouble's sake.")

I also read coverage of Jesse Gray in the *Amsterdam News*. "Rent Strike in Harlem" appeared in *Ebony* in April 1964, and it features many pages of photos of Gray at work at community meetings, in jail, in run-down apartments. The photographer for the *Ebony* story, Don Charles, was arrested along with Gray while photographing a police raid on a rent striker. *The New Yorker* reported on the Harlem rent strike in "The Talk of the Town" section on January 25, 1964, and described Gray's office and his coworker Brown. I read the literature from Gray's campaign for Congress in the clip files of the Schomburg Center for Research in Black Culture, a research center of the New York Public Library at 515 Malcolm X Boulevard, at 135th Street in Harlem. There is also a recording of Jesse Gray's voice at the Schomburg Center; he is on an LP along with Martin Luther King. Pat Hollander's column "The Door's Open" in the *Post,* June 27, 1969, mentioned Jesse Gray's birthplace and his early union work. A paper prepared for the Housing '97 Conference at New York University School of Law Center for Real Estate and Urban Policy and the New York City Rent Guidelines Board on May 14, 1997, entitled "Rent Deregulation in California and Massachusetts: Politics, Policy, and Impacts," by Peter Dreier, said that Jesse Gray's strike inspired tenants movements all over the U.S. and gave birth to the National Tenants' Organization, in 1969. Photos of the rent strikes and of the other tenant movements can be seen at the Web site www.tenant.net, as can Joel Schwartz's essay "Tenant Power in the Liberal City, 1943–1971 (Title I: Challenge and Response, 1949–1963)." I spoke on the telephone with Schwartz, who is a history professor at Montclair State University and the author of "The New York Approach: Robert Moses, Urban Liberals and Redevelopment of the Inner City," and he wondered if the so-called slum clearance plans of Robert Moses did not contribute to the rodent infestation problems in the neighborhoods, given that the construction made for more dirt and debris and, thus, rodent habitat. An article in the *Village Voice*—"By Any Means (Unnecessary)" by Peter Noel, September 1, 1999—said that Jesse Gray was just one of a long line of confrontational politicians that included Malcolm X. I think of him along the lines of Al Sharpton. That's who Randy Dupree, the former city rodent control official whom I mention above, compared him to anyway. When Dupree first moved to New York after serving in the army, he lived in a vermin-infested apartment building that went on a rent strike with the help of Jesse Gray. Dupree, who now lives on Strivers Row, one of the most beautiful streets in Harlem, described his early apartment in Harlem for me thusly: "There were cockroaches all over that place. I remember when people came to the house, I was always embarrassed because there were just cockroaches all over the place. Sometimes, we would cut off the lights and—well, the way that I maintained my sanity was to color them. So what we would do was we would sit there in the kitchen and every time we would see one, we would take a brush and some paint and *blop!* We'd hit them with a color every time one came out. So then, after we painted a bunch of them we'd go shut off the lights and go into the next room and have a drink or something, and then we'd go back into the kitchen and turn on the lights and you'd see all these Technicolor roaches go running all over the place. Like I said, that's the way I maintained my sanity."

## CHAPTER 8: FOOD

Marty Schein and Holmes Orgain's paper "A Preliminary Analysis of Garbage as Food for the Norway Rat" was published in *The American Journal of Tropical Medicine and Hygiene* (vol. 2) in 1953. Information on Schein's life came from *The Mountaineer Spirit,* a newsletter of West Virginia University, where Schein was professor emeritus of biology, and from the May 1999 issue of the newsletter of the Animal Behavior Society, which also described the rest of Martin Schein's career: he studied cattle, in addition to rats and turkeys, and was a highly regarded teacher who, after his retirement, went to Japan with his wife to teach English. "Marty was personally well organized," the newsletter said—he spent his last few years organizing the animal behavior archives at the Smithsonian. Before he died he asked that there be no memorial service, and he donated his body to a medical school. I often wonder if he went with Dave Davis to trap rats in Harlem when Davis was calculating the number of rats in New York City; he is one of those people whom I learned about only after they'd just recently died, leaving me to wish that I could have met them.

## CHAPTER 9: FIGHTS

The quote describing immigrants as "bestial" comes from *Gotham: A History of New York City to 1898* by Edwin G. Burrows and Mike Wallace, a book that I referred to frequently. I first read about Kit Burns and his trial in "Henry Bergh, Kit Burns, and the Sportsmen of New York," an article by Martin and Herbert J. Kaufman, in the *New York Folklore Quarterly* (vol. 28 no. 1, March 1972). (Rat fighting's being replaced as an urban sport by baseball was mentioned in *New York Folklore Quarterly.*) It referred me in turn to numerous contemporary newspaper accounts of Burns, including the *New York Daily Tribune* of September 22, 1868; the *New York World* of November 27, 1866; the *New York Evening Telegram* of Oct 17, 1867; the *New York Herald* of December 24, 1870; and the *New York Sun* of February 24, 1871. Luc Sante's book *Lowlife* also mentions Burns and the rat pit. The letter from Kit Burns to Henry Bergh was in the *New York Herald* and dated September 6, 1868. I read about Henry Bergh in the April 1879 issue of *Scribner's Monthly,* in which appeared "Henry Bergh and His Work," an article by C. C. Buell. "Walking Around in South Street" by Ellen Fletcher, published by the South Street Seaport Museum in association with

Leete's Island Books, talks about the history of Kit Burns's place, which is still there—up until two years ago there was a plaque. Another book that describes a rat fight—"The scene is disgusting beyond description"—is *The Secrets of the Great City,* published in 1868 and written by Edward Winslow Martin; its subtitle is *A Work Descriptive of the Virtues and the Vices, the Mysteries, Miseries and Crimes of New York City,* and it includes a five-panel drawing purporting to illustrate the home range and life expectancy of a young man who migrates to New York City in the 1860s. The captions for the illustrations are as follows: "1. LEAVING HOME FOR NEW YORK 2. IN A FASHIONABLE SALOON AMONGST THE WAITER GIRLS—THE ROAD TO RUIN. 3. DRINKING WITH 'THE FANCY'—IN THE HANDS OF GAMBLERS. 4. MURDERED AND ROBBED BY HIS 'FANCY' COMPANIONS 5. HIS BODY FOUND BY THE HARBOR POLICE." The preface states, "Few living in the great city have any idea of the terrible romance and the hard reality of the lives of two thirds of the inhabitants."

I did not realize that the ASPCA still has officers in the field—they break up dog fights and investigate animal cruelty, among other things—until one day when I was out in the Sunset Park section of Brooklyn, near a live poultry shop, the kind of place rats love, where I ran into an ASPCA officer: I recognized the Henry Bergh-designed shield on the side of his police-car-like vehicle. A nice guy, he knew all about Henry Bergh's battle with Kit Burns, just off the top of his head, and he referred me to the offices of the ASPCA in Manhattan, where I met the office's historian, who gave me a long list of additional newspaper references to stories about Bergh and Burns. While I was at the ASPCA, I learned that when the group had a problem with mice in their offices, it trapped and released them.

## CHAPTER 10: GARBAGE

A blow-by-blow description of the sanitation strike, as seen from the union's side, is in a book published by the Uniformed Sanitationmen's Association titled *Nine Days That Shook New York City*. It includes transcripts of television and radio interviews, documents from the city, state, and federal governments, and pictures of all the players and the situation. It was put together at the time by the public relations consultant working with the union, Howard J. Rubinstein. I also read numerous newspaper stories from the time, including "Explosive Bargainer," which was published in the *Times* on February 3, 1968, and a piece by A. H. Raskin, the great *Times* labor writer, entitled "Mayor and Governor: Knee-Deep in Trouble," published on February 11, 1968. I read contemporary articles as well in the *New York Post* and the *Daily News*. John DeLury's obituary in the *Times,* by Dena Kleiman, was helpful, as was the obituary in the *Post* by Keith Moore, which noted that DeLury said, when he retired, that his greatest accomplishment was in raising the status of "garbagemen" to "sanitationmen." To learn about the fiscal crisis, I read *The Streets Were Paved With Gold* by Joe Klein and *American Metropolis: A History of New York City* by George J. Lankevich. I also read several interviews with Jack Bigel, a consultant to the union at the time; he died in November 2002, and one of his obituaries pointed out that he knew New York City's financial numbers inside and out just the way he knew its municipal and labor leaders. In *Nine Days That Shook New York City,* the U.S.A., to its credit, printed editorials for and against it—and mostly against. An editorial in the *Times* on May 11, 1976, said, "Who runs the city? It's apparently not the Mayor; is it Mr. DeLury? As regards unions, the latest rat-oriented development is the use of giant inflatable rats by unions in front of buildings that the union believes is using work or hiring rules that are against the best interest of the unions." I reported on this phenomenon in a "Talk of the Town" story, first published in *The New Yorker* on August 14, 2000, under the headline "Hot Air: The Man Behind the Rats":

> In New York, there are regular rats and there are giant rats. The regular rats get most of the attention these days; Mayor Giuliani just resurrected his 1997 rat task force, and there is talk of a city rat czar. But it is the giant rats whose population has exploded most visibly in recent months. Since January, two dozen giant rats have arrived in New York. They were sent here from Chicago by one man, Mike O'Connor, the inventor of the giant rat and the president and founder of Big Sky Balloons and Searchlights. O'Connor designs and manufactures the inflatable nylon rats that turn up at the sites of labor actions all over town. "We're—what do you call it—infesting New York with rats," O'Connor said recently, as he drove his pickup truck through the suburbs of Chicago. He had just finished putting up a giant blue King Kong in someone's backyard (in honor of a birthday) and was on his way to another town to install a giant stork (in honor of a birth). "We're doing a lot more storks lately," he said.
>
> O'Connor, a hot-air balloonist from Billings, Montana, founded Big Sky Balloons twenty years ago, after he took an old hot-air balloon, sewed the bottom shut, attached it to a furnace blower, and pumped it full of air to create what he believes to be one of the first commercially produced superpressure balloons—that is, a balloon that stays put instead of rising and drifting away on the wind. Before long, O'Connor moved from balloon-shaped balloons into balloons shaped like just about anything, including bees, dragons, whales, raccoons, owls, witches, elephants, and, on behalf of the Chicago Bulls, giant bulls.
>
> O'Connor designed his first giant rat in 1987, for a construction union in Chicago. "They wanted a really mean one," O'Connor recalled. "So I drew one up for them. I was the artist on that one. And I showed it to them, and they said, 'No. Not mean enough.' And I said, 'You mean like a really rabid rat?' So I made it really mean, and they saw it and said, 'That's it!' "
>
> Since then, O'Connor has sold close to a hundred giant rats—seventy-five of them in New York. They come in three sizes: twelve, fifteen, and twenty-five feet. The twelve- and fifteen-footers are the most common, and they sell for $3,775 and $5,025, respectively. The twenty-five-footer goes for $7,225. The design of each rat is essentially the same. "We're getting into changing them a little," O'Connor said. "You know, with festering-looking nipples on their chests. We're getting into airbrushing."

Lately, amid the renewed vigor of the labor movement, sales of O'Connor's giant rats have taken off. "We sell four or five a week," he said. O'Connor sells his rats exclusively to labor unions, which use them as attention grabbers, though he once sold a giant cat to a bank that was the target of some demonstrators who had put up one of his rats. "It was actually a cougar, but it looked like a cat," O'Connor said. "It towered over the rat. It was pretty cool. But it was all done in good taste."

Recently, one of O'Connor's giant rats presided over a labor demonstration on Broadway near Columbus Circle. It stood behind police barricades, fifteen feet tall, its claws trembling in the summer breeze. One of the rat's caretakers was Mickey Whelan, a member of Local 608 of the United Brotherhood of Carpenters and Joiners of America. "This is just standard. There's no big deal about this one," Whelan said. He had an Irish brogue and a handshake like a two-by-four. "It's not to say that the guys on this job are rat workers or anything. Far from it. It's to say that the workers here are getting exploited."

Just then, a skinny young man who looked to be in his late twenties stopped next to Whelan. He was carrying a large portfolio of artwork. "Nice rat!" he said.

Whelan, his arms folded across his chest, eyed the young man skeptically. "Thanks," he said.

"Have you seen our rat?" the man asked.

"*Your* rat?" Whelan said.

"Yeah, our rat's over there." The young man pointed across town. "We're on strike over at the Museum of Modern Art."

"Oh, all right then," Whelan said.

"Yeah, we got a rat, too," the young man said again. He was pretty excited. He tried to high-five Whelan, but Whelan didn't notice. Instead, Whelan reached out to shake his hand.

"Well, good luck, then, brother," Whelan said.

New developments in the inflatable-rat area include the use of giant inflatable cats by landlords.

The bar in which I talked to John DeLury about his grandfather was Waterfront Ale House on Atlantic Avenue in Brooklyn.

## CHAPTER 11: EXTERMINATORS

The first professional exterminators in America are mentioned in *The Ratcatcher's Child,* a history of the American pest control industry written in 1983 by Robert Snetsinger, a professor of entomology at Pennsylvania State University. Snetsinger recounts the transition of the title *exterminator* to *pest control operator,* when the exterminators association was under the leadership of William Buettner, who was considered an especially strong leader: through his efforts, exterminators were recognized as essential during wartime, and the only other workers in the service industry to gain such a status were undertakers. Frederick Wegner's career was described in the *Times* on September

3, 1893; August 8, 1893; and August 11, 1893. (Wegner also killed cats, prowling through the Central Park Zoo at night with a rifle. "In this way he managed to combine his business of catching rats with the pleasure of shooting cats," a report said.) Harry Jennings's obituary said, "Harry Jennings was the foe of all rats and vermin." Juan Colon is profiled in "The Lonely Soldier of Extermination" in the *Times*, August 13, 1995. I interviewed John Murphy, and his quote, which appears on the epigraph page, came from an interview in *Pest Control* that appeared on October 1, 2001, in an article entitled "Bootleggers & Rats: This Rat Call Took Inspection to a Whole Different Level." The race to out-alphabetize rival pest control businesses may have been begun by one of the oldest New York–area exterminators still in existence, Abelene Pest Control. According to Snetsinger, Abelene was begun in New York City in 1927 by Walter O. Blank, a German who exterminated the rats that overran the trenches during World War I. He had heard that abalone oil worked as an insecticide, but he preferred the name *abelene* and figured that a firm so named would rank first in alphabetical exterminators listings. The inter-pest-control-firm alphabetization competition still characterizes the industry as it appears in phone directories today. For example, New York City listings for pest control firms include AAA Absolute, AAA Advanced, AAA Affordable, followed closely by Attack, Ban-the-bug, Bug-Off, Bull Dog, and Bustabug Pest Control. (There are no pest control firms with names that begin with the letter *Z*, though a few begin with letters from the end of the alphabet, such as Swat, and Victory, and a firm called The End Is Near.)

Information on the beginning of the city's exterminating force came from health department literature, and the number of people living in housing projects came from the New York City Housing Authority. According to the authority, the number of people who live in housing projects in New York City exceeds the populations of such cities as Atlanta, Minneapolis, and Miami. In 2002, 51.6 percent of the residents of housing projects were listed as black; 43.5 percent as Hispanic or other ethnic group such as Asians or Native Americans; 4.9 percent as white.

## CHAPTER 12: EXCELLENT

I first read about Milwaukee's expertise in rat fighting in a federal report entitled "The Relationship of Solid Waste Storage Practices in the Inner City to the Incidence of Rat Infestations and Fires" by Robert M. Wolcott and Burnell W. Vincent, which was published by the Environmental Protection Agency in the mid-seventies. Dave Davis applauds Milwaukee's rodent control in an article in the *Milwaukee Journal* entitled "Rat Program of City Praised," which appeared on March 30, 1968. In 1971, Milwaukee's rat control program was rated among the best of eleven cities surveyed by the federal department of Health, Education, and Welfare, according to the *Journal* of September 11, 1971 (the early edition of the paper accidentally referred to the program praised as a rent control program). The quote on the monument at the Wisconsin Workers Memorial was originally taken from a book called *The Rise of Labor and Wisconsin's Little New Deal.* That Milwaukee lost some sixty thousand jobs in the recession between 1979 and 1982 came from the *Encyclopaedia Britannica.* When I met with Don Schaewe, he demonstrated an experimental rat-hole-activity monitoring technique—crumpling a ball of paper and stuffing it into a rat hole and returning the next day to see if the paper has been expelled. This technique was first suggested to Don Schaewe by Bobby Corrigan.

Articles I read describing the career of John Norquist, Milwaukee's mayor at the time of my visit, include "Was It Harassment by Mayor or Sex Scandal?" by De Wayne Wickham in *U.S.A. Today,* on January 13, 2001, and "Scandals Begin to Tarnish Wisconsin's Political Luster" in *The New York Times*, on July 9, 2002. Norquist was quoted by the Associated Press on April 22, 2002, as saying, "I made a mistake, as I have said before, and I accepted responsibility for that." The Associated Press also ran a story about Alderwoman Rosa Cameron pleading guilty to funneling federal grant money into her campaign on December 27, 2002. Other articles I read in the *Journal* about Milwaukee included "Krumbiegel Asks Help in Rat War," which appeared on November 11, 1966, and "City Begins Showing Rat Control Films," which ran on June 15, 1971. A photo of Ramon Hernandez dressed as a giant rat was in the *Milwaukee Journal* on June 18, 1971. A study by a group called Erase Racism, reported in the *Times* on June 5, 2002, showed that Milwaukee was one of the most segregated cities in the United States; according to the study, 82 percent of the population of blacks would have to move to be evenly dispersed among the population. Detroit was at 85 percent, Chicago at 81 percent. The nation's one hundred largest metropolitan areas average 60 percent. The most segregated suburban areas in the country are New York City's suburbs on Long Island, at 74 percent. "African-Americans have faced isolation far more than any other group, especially on Long Island," said one of the consultants who analyzed the segregation patterns. "It's almost like a township in the South African sense," another expert said.

I read a story on Bobby Corrigan that appeared in a special issue of *Pest Control Technology* called "Leadership Winners 2000." I also read, as I mentioned above, his book, *Rodent Control*, and many of his columns and articles, and naturally I took copious notes during his lectures. Nearly all the pest control operators at the conference had nothing but praise for him or for some advice he had offered them at some point in their career. "Bobby's the greatest" (or slight variations of that phrase) was something I heard countless times at the Courtyard Marriott. I did not learn that Corrigan is a poet until after I left the conference. Here is a poem, used by permission of Corrigan, that was originally published by *Pest Control Technology*; it is entitled "5:41":

At 5:41
my yard
is a diffused wash of a yellow
that I've never seen here,
or anywhere, before.
But the sun
is not yet even visible.

How many other colors
and mystical rarities

of the dawn,
the deep night,
the busy mid day

do we sleep or work through?
Only once here,
then forever in our lives gone.

Concerning William Jackson's early cat-versus-rat work, which I mention in reference to Jackson's talk in Chicago, I would add that in 1986 Jamie Childs conducted research on city cats and rats and found that cats will catch only juvenile or subadult rats and do not complete their chases of adult rats. "Although adult cats and larger rats were frequently observed in close proximity, no aggressive behavior was directed by rats toward cats, and generally these species coexist peacefully in alleys," Childs observed. An article that inspired me to consider rat trapping in the first place is Child's "And the Cat Shall Lie Down With the Rat," published in *Natural History* in June 1991. William Jackson also investigated the rats in the nuclear tests and wrote many papers on their survival, as well as on rodenticide resistance. (The first case of resistance was reported in Scotland, in an article in *Nature,* on November 5, 1960, by C. Mary Boyle: "A Case of Apparent Resistance of *Rattus norvegicus* Berkenhout to Anticoagulant Poisons.") In Chicago, I asked Jackson what had happened to the idea, first floated in the seventies, of introducing sterilized male rats into the rat population, in an attempt to bring down reproduction rates, and he said that strategy hadn't panned out. And in Chicago I also talked to Stephen Franz over lunch about his wild rat colony; the rats apparently lived in a building that is now an antique store on the outskirts of Albany.

## CHAPTER 14: PLAGUE

The Black Death is, of course, the subject of so many books that, when I read about the disease, I felt a little like a flea on the back of the rat that is the study and history of the bubonic plague. My reading included Philip Ziegler's *The Black Death,* and a book that is composed of several lectures by David Herlihy, *The Black Death and the Transformation of the West.* Rosemary Horrox is the editor and translator of *The Black Death,* a vast collection of medieval texts that have to do with the plague; it makes for a kind of casual Black Death reading that can be perversely satisfying. I also read *Plague! The Shocking Story of a Dread Disease in America Today* by Charles T. Gregg, a book, published in 1978, that isn't as hyperbolic as it sounds. Many of the descriptions of natural phenomena that were thought to lead up to a plague epidemic came from *The Black Death: A Chronicle of the Plague,* compiled by Johannes Nohl from contemporary sources, translated by C. H. Clarke, Ph.D, with numerous illustrations (London: George Allen & Unwin Ltd., 1926). Articles I read included "How a Mysterious Disease Laid Low Europe's Masses" by C. L. Mee Jr., which was published in *Smithsonian* in February 1990; this article describes a flea as being the size of a letter *o*. *Fighting the Plague in Seventeenth-Century Italy* is by Carlo M. Cipolla, who was an economic historian at the University of California, Berkeley, and at universities in Venice, Turin, and Pavia. I also read *Miasmas and Disease* by Cipolla and a treatise called *The Basic Laws of Human Stupidity,* which was a best-seller in Italy and eventually made into a play.

I talked to Bruce Colvin about rats' general disease-spreading capability. "Rats are very capable of elevating bacteria in our environment because they live in sewers and back alleys and search for food in the gutter," he said. I once interviewed Professor Glass, who is at the John Hopkins Bloomberg School of Public Health, for a magazine piece that I was writing about rats, and he notes, "They have an awful lot of stuff that they carry." Rats and plague are also covered in chapter 23 of the *Textbook of Military Medicine: Medical Aspects of Chemical and Biological Warfare,* as published on-line by the Virtual Naval Hospital, at www.vnh.org. Anthony Burgess's analysis of the city as hero is in the introduction to the Penguin Classics edition of *A Journal of the Plague Year.*

## CHAPTER 16: PLAGUE IN AMERICA

I learned about Yersin's discovery of the plague bacillus from Gregg's book *Plague!* (Shibasaburo Kitasato, a Japanese microbiologist, identified the plague bacterium at the same time, though his findings have long been considered controversial.) I also read an article by Ludwik Gross entitled "How the Plague Bacillus and Its Transmission Through Fleas Were Discovered: Reminiscences from My Years at the Pasteur Institute in Paris," in the August 1995 *Proceedings of the National Academy of Science* (vol. 92, pp. 7609–11). Ludwik Gross worked at the Pasteur Institute in the forties and then moved to the United States, where he worked at the Bronx Veterans Administration Hospital until his death in 1999. He enjoyed writing to scientists and asking about their discoveries; he collected their responses. Another story Gross told about the plague involved Edmond Dujardin-Beaumetz, a friend of Yersin's who worked in the plague laboratory when Gross visited the Pasteur Institute as a young guest investigator. The story goes like this: "Dujardin-Beaumetz showed me a tube filled with live bacilli of plague. He told me that not only humans and rats but also monkeys, guinea pigs, mice, and many other species are susceptible to the plague bacillus. But not the chicken. Among the species resistant to the plague is the chicken. 'Look at this tube full of live bacilli of the plague,' said Dujardin-Beaumetz to me, taking out of a cabinet a small tube marked with a red pencil *B.P.* [which was short for bubonic plague]. 'This small tube contains sufficient quantity of live plague bacilli to infect and kill the population of an entire district of Paris,' he continued. 'We injected a similar quantity of live bacilli into the peritoneal cavity of a young chicken in our laboratory,' Dujardin-Beaumetz told me, 'and the chicken remained in good health. In fact, the next day she laid an egg. Surprisingly, the chicken got lost, presumably flew out of a small open window in the adjoining laboratory. We were frantic and looked for this animal all over, afraid that it may spread the deadly disease, but we could not find the chicken. Only several days later did we learn that the chicken was caught by a house superintendent, residing on a street adjoining the Institute, on rue Falguiere. Not realizing the chicken came from our laboratory, he roasted the chicken and consumed it, sharing the unexpected meal with his family. The plague bacilli were presumably destroyed by roasting the chicken. Nothing happened to them. They all remained alive and well.' "

I read about the plague in San Francisco in *Plague!* but, mostly, in two long articles: "The Black Death in Chinatown: Plague and Politics in San Francisco, 1900–1904" by Philip Kalilsch, in *Arizona and the West,* and " 'A Long Pull, a Strong Pull, and All Together': San Francisco and Bubonic Plague, 1907–1908" by Guenter B. Risse, in the *Bulletin of Medical History,* 1992 (vol. 66, pp. 260–86). Honolulu's burning is discussed in "Plague on Our Shores," a series by Burl Burlingame that appeared in the *Honolulu Star-Bulletin* on January 25, 2000. I read *The Barbary Plague: The Black Death in Victorian San Francisco*, by Marilyn Chase, after I wrote about San Francisco's plague. I wish I'd read it before, given its documentation of the differences between Blue and Kinyoun (and of the bitter and racist remarks Kinyoun made after he was kicked out of San Francisco); it's worth getting just for the pictures of the debonair Blue versus the arrogant and stodgy-looking Kinyoun, and the photos of the "rattery" where scientists looking for plague dissected rats. One historian has argued that San Francisco's first plague epidemic merely showed the world what was already there: "It revealed the best and the worst in people, and the lower and higher motives that businessmen, newspaper editors, politicians, and physicians normally repress. The result was a situation so tense that the true villains and real heroes quickly emerged from the placid fabric of routine city life."

## CHAPTER 17: CATCHING

When I went to the Bushwick section of Brooklyn, I read about the history of the neighborhood in *The Neighborhoods of Brooklyn*, a book edited by Kenneth Jackson and John B. Manbeck and published in 1998 by the Citizens Committee for New York City and Yale University Press. I also read an article in the *Village Voice*, "Close-Up on Bushwick," December 27, 2002. Some of the plants were in various plant and nature guides, including *Urban Wildlife* by Sarah Landry in the Peterson First Guides series, 1994. The study to which Ann Li refers to when she is talking about rats sensing each other's stresses and fears is called "Regulation of Ovulation by Human Pheromones," written by Martha McClintock and Kathleen Stern for the March 12, 1998, issue of *Nature*. I learned about the times that plague came to New York City, or nearly came, in newspaper articles such as "Fear the Plague Is Here," which was in the *Times* on November 19, 1899. I also read about the *Taylor* in Gregg's book *Plague!* "The Plague Ship's Cargo," which ran on November 28, 1899, discusses the shipment of coffee that was on the *Taylor*. During the ship's quarantine, the coffee was apparently allowed to sit on the piers and eventually continued on to the coffee merchant who sold some in New York and some in Chicago, which concerned people but ended up not being a problem, according to a May 22, 1900, *Times* story, "The 'Plague' Coffee Has Been Consumed." Other plague reports include "Liner Held Back by Plague Order; Passengers from Havana May Be Detained for Seven Days," which was published on July 14, 1912, and describes fears over plague infection from a plague outbreak in Puerto Rico at the time—they were afraid to let sick passengers off the boat—and "Signs of Bubonic Plague in Three American Cities," which was in the *Times* on February 8, 1925. A engrossing account of bubonic plague in New York is included in Joseph Mitchell's story "The Rats on the Waterfront," one of the best pieces of rat reportage ever, which was published in *The New Yorker* as "Thirty-Two Rats from Casablanca," on April 29, 1944, then included in the book *The Bottom of the Harbor* under the title "The Rats on the Waterfront." An article about the rat fighters in port—"Ship Rats a Minor Problem Here but Every Vessel Gets a Once-Over"—ran in the *Times* on Feb 3, 1952. More recently, the *Daily News* reported on the man who caught plague in the Southwest and nearly died of it while in New York in an article entitled "Plague Costs Man His Feet," January 17, 2003.

The information about General Ishii and Japan's biological weapons program came from a speech by Judith Miller, a *Times* reporter who has covered biological weapons extensively, that was given at the University of Virginia's Miller Center for Public Affairs and subsequently reprinted in a university publication, *Miller Center* (fall 2001), that was posted on-line, and from the Virtual Naval Hospital's *Textbook of Military Medicine: Medical Aspects of Chemical and Biological Warfare*. Mostly, I consulted *The Biology of Doom: The History of America's Secret Germ Warfare Project* by Ed Regis, which I first saw mentioned in *The New York Times Book Review* of January 23, 2000, in an review by Timothy Naftali entitled "Death Factories: A History of Germ Warfare and America's Involvement in It." In addition to all kinds of germ warfare experiments conducted on an unwitting American public, *The Biology of Doom* describes the government agents carrying germ-infused lightbulbs and dropping them in between subway cars and, subsequently, measuring their effect with an air sampler that they called a Mighty Mite. Something that upset me almost as much as secret government agents releasing germs on subways is a secret agent bragging about lying to a man sitting next to him on the subway. "While riding train to 23rd Street Station, a man asked me where I got the nice little plastic case," an agent wrote. "I told him all the hardware stores over town had them. He is going to buy one."

## CHAPTER 19: A GOLDEN HILL

In further investigating the history and former geography of Edens Alley, I used a number of different sources. At the city archives I looked at all the plans for adjustments and additions to all of the buildings on the alley; I saw John DeLury's signature, for instance, on the paperwork required to add on to the U.S.A.'s building in the sixties. In December 1933, the *New York Herald Tribune* ran a story entitled

"Gold St., Refinery Center Here, Traces Name Not to the Metal but to Colonial Wheat Field." I found an undated clip with the blocky-type look of something from an old *Times*—I can't tell for certain—that is entitled "March on With Time: Signs of Last Century Still Persist In and Near the 'Swamp' Area Downtown." An amazing book about the area of lower Manhattan that once covered the swamp and is now public housing and office towers is *The Destruction of Lower Manhattan* by Danny Lyon. It includes photographs and journal entries from when the Swamp was being demolished (as well as photos of the portion of Washington Street that was demolished when the World Trade Center was constructed). "In 1967 over sixty acres of buildings of Lower Manhattan were demolished," Lyon writes. The possible translation of the word *Manahachtanienk* as "the island where all became intoxicated" comes from the book *Native American Place Names in New York City* by Robert Steven Grumet, published by the Museum of the City of New York in 1981.

Some books that were helpful in detailing the career of Isaac Sears and explaining his times include *A Mighty Empire: The Origins of the American Revolution*, by Marc Egnal; *The Story of American Freedom* by Eric Foner; *Farewell to Old England: New York in Revolution* by Ellen F. Rosebrook; *Divided Loyalties: How the American Revolution Came to New York* by Richard M. Ketchum; *The American Revolution* by Edward Countryman; and *The Battle for New York: The City at the Heart of the American Revolution* by Barnet Schecter. Pauline Maier's book *The Old Revolutionaries: Political Lives in the Age of Samuel Adams* looks at the lives of several colonial Americans who were in the generation prior to the founding fathers, the group within which the first flames of revolution were born; it features a chapter on Sears. Maier's book *From Resistance to Revolution: Colonial Radicals and the Development of American Opposition to Britain, 1765–1776* explains the times from which Sears and his compatriots emerged. The only book ever written on Sears is a dissertation that was about to become a book but didn't and was written by Robert Christen in 1968; it is entitled "King Sears: Politician and Patriot in a Decade of Revolution." Christen himself was president of the New York City Board of Education from 1976 to 1977; he was a lifelong resident of the Bronx and died when he was 53. At Manhattan College, where he taught before working for the school board, he was a well-known anti–Vietnam War protester, a cofounder of Manhattan College's Peace Institute, and creator of "The Anatomy of Peace," which has been described as one of America's first courses in peace studies. I learned about tavern etiquette—pipe and mug sharing, the cutoff of drinks to people not with the majority's cause—while on a visit to Fraunces Tavern, a tavern left over from revolutionary New York, with my daughter and her first-grade class.

For detailing the Battle of Golden Hill, I relied greatly on Christen's work and also on *Labor and the American Revolution* by Philip Foner, which describes the lead-up to the Battle of Golden Hill. Foner points out that the Liberty Boys weren't always interested in all the lower classes' working conditions; they were clannish. "This movement will not be romanticized for in some communities, the Sons of Liberty were indifferent to the plight of other lower-class elements in the struggle," Foner wrote. Foner is the historian who said that you could write a book that just listed all the derogatory names people had come up with for the lower classes. Foner also wrote *History of the Labor Movement in the United States,* and in the section entitled

"Labor and the American Revolution," he reports that there was a Sons of Liberty chapter in England ("You have only to preserve, and you will preserve your own liberties and England's too," they said) and in Ireland, where they assisted the Americans financially, recruited Irishmen to fight in George Washington's army, and drank toasts declaring, "Sons of Liberty throughout the world!" In Boston, workers who were members of the Sons of Liberty refused to work for people whom Paul Revere described as "enemies to this country," so that when the British wanted to build barracks for troops in Boston, they were forced to send for workers from out of town—i.e., from New York.

Christen refers to the Battle of Golden Hill as the first move toward concerted physical resistance in the American Revolution. You can see the broadside that Brutus wrote that so angered the British troops at the New-York Historical Society. What was called the Fly Market is where Louise Nevelson Plaza is today, on Broad Street. A helpful guide to an older New York is the *Historical Guide to the City of New York,* which was compiled by Frank Bergen Kelley "From Original Observations and Contributions Made by Members and Friends of The City History Club of New York" and published in 1909. That Sears and his friends expected to be paid came from *The Old Revolutionaries*. Maier wrote of "the luxury of unrewarded patriotic service" and added this: "Confident in the material implications of liberty, New Yorkers talked less of virtue than of interest, and turned naturally to the language of business in the business of revolution: 'Stocks have risen in favor of Liberty,' Alexander McDougall wrote Samuel Adams in June, 1774."

Fortunately, I'm not the only one who has noticed Golden Hill even after it has become barely noticeable. After I realized where it was, I went back to Hemstreet's 1899 book, *Nooks and Corners of Old New York,* and read this: "Golden Hill, celebrated since the time of the Dutch, is still to be seen on the high ground around Cliff and Gold Streets. Pearl Street near John shows a sweeping curve where it circles around the hill's base, and the same sort of curve is seen in Maiden Lane on the south and Fulton Street on the north." The details on the disappearance of the marker at the site of the battle came from old newspapers. One said, "Many have been the complaints against New York because of its apathy in respect to its own history, rich as it is in interesting incidents." An article I cited about the disappearance of the plaque marking the site of the Battle of Golden Hill was "Historic Tablet on a Pilgrimage," from the *Times* on March 20, 1910. The article describing the last Liberty Pole replacement is from the *Times* on August 30, 1952—"Saw Brings Down City Liberty Pole: Decayed Staff Latest to Fall in a Long Historic Line but New One Will Rise." The inscription on the plaque itself was printed in the *Times* on September 17, 1898, and I saw a photograph of it in the Fraunces Tavern museum. Robert Moses was the official responsible for cutting the Liberty Pole down. Near the site of the Liberty Pole, in today's City Hall Park, where I only recently saw a gathering of cabdrivers protesting tax license changes, there is a plaque in honor of Debs Myers; it is one of my favorite plaques in the city and says, "Do the right thing and nine times out of ten it turns out to be the right thing politically." I feel strongly that there should be a plaque marking Golden Hill. A plaque may not sound like much given the trend toward audiovisual and computer-driven historical tools that are "interactive," but a plaque marks a place, which is important. I read about the plaque that my father saw in an article entitled "Hercules Mulligan, Secret Agent," which ran in *Daughters of the American Revolution Magazine* in March 1971. The plaque on which my father read about Hercules Mulligan said, in part:

> BORN DERRY, IRELAND 1740
> DIED NEW YORK CITY 1825
> DESPISED BY COMPATRIOTS FOR CONSORTING WITH THE BRITISH, MULLIGAN SILENTLY PERSEVERED UNTIL NOVEMBER 25, 1783 WHEN GEORGE WASHINGTON LED HIS VICTORIOUS CONTINENTAL ARMY INTO NEW YORK AND BREAKFASTED HERE WITH MULLIGAN AND HIS FAMILY.

Until recently the plaque was in the vestibule of 160 Water Street. Now, I don't know where it has gone.

# ACKNOWLEDGMENTS

Gillian Blake; Dena Rosenberg; Greg Villepique; Sara Mercurio; Karen Rinaldi; Steven Boldt; Sarah Chalfant; Tim Farrington; Andrew Wylie; Rachel Sussman; Zoe Pagnamenta; Anna Wintour; Jay Fielden; Joanne Chen; Jill Demling; Bess Rattray; Sally Singer; Irini Arakas; Laurie Jones; Amy Astley; Hugo Lindgren; Gerald Marzorati; William Jackson; George Ladd; John Murphy; Robert Corrigan; Walter Schroeder; Stan Friedman; Greg Zinman; Nick Paumgarten; Marshall Heyman; Susan Morrison; Eric Etheridge; Christophe Barbier; BookCourt; Jessie Graham and Dan Segall; Weiden & Kennedy; Barbara Brousal; Javier Acevedo; Denys Sandoval; Edwyn Smith; Gustavo Camopos; Manny Howard; Lauren Collins; Christopher Mellon; Marty Skoble; Meg Lamason; the Map Division of the New York Public Library; Hastings-on-Hudson Public Library; New York City Archives; Oregon Health and Science University; National Endowment for the Arts; Stephen L. Zawistowski; Russell Enscore; Multnomah County Library; Greg Radich and Camille Schaewe; Jack Conley; Satoru Igarashi; Steve Miller; Paula Grief Zanes and Dan Zanes; Jennifer Marshall and Andrew Mockler; Jim Leinfelder; Michael Thomas; Skip McPherson and Son, Inc.; Kassie Schwann and Brian Rose; Mia and David Diehl; Maureen Harrington; Maggie and Charlie Sullivan; Linda and Donald Desimini; Josh Cole; Peter Scotch; Debbie and Tom Quinn; Kathy and Peter Quinn; Bill and Kristen Sullivan; Matthew "Matt" Sharpe; Jill Desimini and Dan Bauer; Mary Elizabeth and Robert E. Sullivan; Samuel Emmet and Louise Grace.

## A NOTE ON THE AUTHOR

Robert Sullivan is the author of *The Meadowlands* and *A Whale Hunt,* both *New York Times* Notable Books of the Year, and a recipient of a National Endowment for the Arts creative writing fellowship. A contributing editor to *Vogue,* he is a frequent contributor to the *New Yorker.* His work has also appeared in *Condé Nast Traveler,* the *New York Times Magazine*, and the *Oregonian*. He lives in Hastings-on-Hudson, New York.

## A NOTE ON THE TYPE

The text of this book is set in Bembo. This type was first used in 1495 by the Venetian printer Aldus Manutius for Cardinal Bembo's *De Aetna*, and was cut for Manutius by Francesco Griffo. It was one of the types used by Claude Garamond (1480–1561) as a model for his Romain de L'Université, and so it was the forerunner of what became standard European type for the following two centuries. Its modern form follows the original types and was designed for Monotype in 1929.